T0296159

CAMBRIDGE BIOLOGICAL STUDIES

General Editor: C. H. WADDINGTON

FORM AND CAUSALITY

IN

EARLY DEVELOPMENT

FORM AND CAUSALITY
IN
EARLY DEVELOPMENT

by

ALBERT M. DALCQ

Professor of Human Anatomy and Embryology at the
University of Brussels

CAMBRIDGE
AT THE UNIVERSITY PRESS
1938

CAMBRIDGE
UNIVERSITY PRESS

University Printing House, Cambridge CB2 8BS, United Kingdom

Published in the United States of America by Cambridge University Press, New York

Cambridge University Press is part of the University of Cambridge.

It furthers the University's mission by disseminating knowledge in the pursuit of education, learning and research at the highest international levels of excellence.

www.cambridge.org
Information on this title: www.cambridge.org/9781107654488

© Cambridge University Press 1938

First published 1938
First paperback edition 2014

A catalogue record for this publication is available from the British Library

ISBN 978-1-107-65448-8 Paperback

CONTENTS

Figures 7 and 33 are printed as plates

FOREWORD

This contribution to synthetic Embryology is neither a text-book nor a treatise. The author does not pretend to master in a few pages the huge amount of facts and ideas concerning the early stages of development. He thinks it necessary to select the data which appear most significant at the present time for a general comprehension of the ontogenetic problem. He hopes his work will be useful to young biologists who feel the holy enthusiasm for research and are anxious to decide in which way they could best direct their efforts.

A. M. D.

30 *April* 1938

CHAPTER I

The recent embryological movement and the scope of this book

Embryology may seem, at first sight, a kind of Penelope's web. From the end of the eighteenth century, and especially from the middle of the last one, the same eggs of various species have been again and again minutely examined by research workers who have devoted themselves to the enigma of development.

In spite of the enormous amount of data which have in this way been heaped up in an abundant literature, it is always necessary to take up again and again the study of the same materials. Not that the earlier observations are often erroneous, far from it. But ontogenesis is one of the processes of life which is most difficult to integrate in the usual frames of our mentality. With its innumerable problems connected with the physical, chemical, cytological and even philosophical disciplines, it requires a mind specially trained and ever ready to adapt itself to unforeseen facts. The omission or the misinterpretation of the slightest detail may entirely change the significance of a result and thorough reinvestigations are therefore often necessary.

The evolution of descriptive Embryology is, in this respect, characteristic. At the time of the Great War one could, it seemed, consider this science as fully studied. For Invertebrates, as well as for Vertebrates, a series of instructive monographs described accurately, in animals quite representative of each group, the phases of development. In 1919, the late Professor Albert Brachet, under whose guidance it was my privilege to study, was finishing his well-known *Traité d'Embryologie des Vertébrés*. I often heard him express the hope of having built on solid ground and consequently of having achieved a lasting work. After the premature death of that great scientist, my friend Prof. Pol Gérard and myself had to review that treatise for its second edition. We were compelled to modify completely nearly a third of the book (1935), and it now appears that we were not sufficiently drastic. Concerning Invertebrates, the descriptive progress seems less rapid. It is, however, characteristic that, in all species which have been submitted to attentive experimental researches, the authors of these had to amend, correct or complete the classical descriptions of normal ontogenesis.

In experimental matters, the evolution of knowledge is still more rapid. What a magnificent record is that of the achievements of causal Embryology in these last twenty-five years! Although I was then making my first steps in research work, I remember very exactly the ideas that were current, just before the War, in Brachet's laboratory. Without any doubt, the future success of "*Entwicklungsmechanik*" was an article of faith, but Brachet preferred christening the young science "causal Embryology". Certainly, also, the facts of regulation in the first blastomeres, when isolated, of the sea-urchin, the *Amphioxus* and the newt were never lost sight of. The theoretical importance of the conception of "germinal localization", then recently expressed by E. B. Wilson, was, however, often emphasized in opposition to Driesch's discovery. But it was vainly attempted, for the frog's egg, to harmonize the results obtained by destruction of the first blastomeres—and exactly understood, thanks to Brachet himself—with the regulation after reversing the germ, i.e. the experiment of Schultze. In the newt, we did not at all suspect the relation between the processes of regulation in the young stages and the lens-inducing actions of the optic vesicle, demonstrated at that very time by Spemann's epoch-making work. We had, however, the feeling that a new and concrete value was thereby given to the conceptions of self- and dependent-differentiation, a legacy of Roux's prescience. In Fishes, the lesions made by Kopsch at certain places of the blastoporal lip helped to anchor us to the fallacious theory of concrescence. In Birds, defects had been obtained by the same author and American investigators; but the results of these experiments did not at all explain to us the meaning of the primitive streak. This enigmatic feature of early development was erroneously thought by us to be common to all Amniotes, including Reptiles, and we attempted in vain to guess its relation with the gastrulation of Fishes and Amphibians. A fact worthy of special mention is that it became for the first time possible, in those same years 1913–14, to experiment on Mammalian eggs: we wondered at Brachet's successful cultivations *in vitro* of the young blastocysts of the rabbit. Ascidians looked to us, of course, a very strange material: we knew of the definite defects obtained in their case, but this seemed to us really singular. The sea-urchin egg interested us, but more by the then thrilling analysis of its fertilization than by its morphogenesis. We had heard, however, some rumour of the theory of Neo-vitalism. It was generally not sympathetic to our young minds. We attempted to refute it on the grounds that regulation was abnormal and less important than results supporting the mosaic theory.

Very incomplete, indeed, is this sketch of the thoughts prevailing, twenty-five years ago, in a laboratory devoted to causal Embryology. It allows us, nevertheless, to appreciate the progress which has now been made in many directions. Our knowledge, at that time, was altogether rather incoherent and extraordinarily rich in promises. Each of the subjects I have mentioned has been the source of continuous progress and remarkable achievements. The most varied methods have been applied to an ever-increasing number of species. Between the nearly contradictory observations of predictable deficiencies and astonishing regulations a bridge has been thrown by the study of morphogenetic functions. The old struggle between Vitalism and Mechanism has lost much of its acuteness. It has been recognized that it is unnecessary to decide *a priori* if the intimate processes of life are or are not resolvable in terms of our actual physical and chemical knowledge. The problems posed by early development have been faced with a complete spirit of objectivity, a philosophical climate which may be styled a pragmatic Organicism.[1]

A large part of this renovation has been due to the discovery of embryonic inductions. The new order of ideas introduced and methodically developed by Spemann's school had been marvellously productive. Avoiding any premature physico-chemical thesis, those investigators have, beyond anything else, accurately registered the answers of the germinal system to logically formulated questions. Realizing organicism in its truest meaning, they have in a few years accumulated a quantity of data, the remarkable

---

[1] Concerning the philosophical interpretations of development, cf. Russel, 1930; Bertalanffy and Woodger, 1933.

outcome of which is plainly established by the recent book of their leader. The American school has not failed to play an indirect but very significant part in this movement. The fine analysis of the relations between the development of the limb bud and the whole embryo, as performed by R. G. Harrison and his distinguished co-workers, has an importance which must at least be emphasized by this very brief allusion. This post-War movement of causal Embryology has taken a rapid extension. Most Chordate germs, those of *Amphioxus* and Tunicates, of Fishes, Reptiles,[1] Birds, and even Mammals, have been examined according to the principles introduced for the Amphibians. In spite of the apparently quite different embryonic forms encountered in Echinoderms, Molluscs, Worms and Insects, skilful and tenacious workers have disclosed to us, at least in its general features, the germinal organization of those Invertebrates.

Such achievements could not fail to attract the attention of embryologists inclined to synthesis. The results gathered up to the end of 1929 have been summarized by Schleip in an incomparable treatise. The extensive textbook of J. S. Huxley and G. R. de Beer, *The Elements of Experimental Embryology*, has discussed primordial ontogeny with the most scrupulous attention. We also have Spemann's memorable book, which was, it must be noticed, already written, for its main part, in 1931, the time of his "Silliman lectures". These excellent contributions are, however, far from making unnecessary an attempt conceived from a rather different angle. In spite of recent advances in causal analysis, the barrier between the normal and experimental data is not overcome in current conceptions. It is of particular difficulty in the cases where *parts of the germ have their fate thoroughly changed with the apparent tendency to restore the normal structure of the whole.* Such processes of *embryonic regulation* may be now considered as of general occurrence. The sole scientific attitude regarding them is to search for an organization of the egg that can account for all results, including the regulation data presented by normal, operated or altered eggs. Normal development (*normogenesis*), anomalies artificially produced (*paragenesis*) and the apparent effort of the germ to build up a normal embryo,

[1] Reptiles remain, however, from the experimental viewpoint, an unknown province.

in spite of severe amputations (*regulation*), must receive one common solution. We must know how much such a fundamental organization differs between species of various Classes, Orders, Phyla. We want to understand how it acts from the awakening of the egg by fertilization or activation until the moment where organs are formed and acquire their histological differentiation. This programme is of course an ideal, which can only be progressively fulfilled. I hope to convince my readers that, thanks to the reciprocal illumination of recent descriptive and experimental results, the causality of animal form, which seemed beyond the goal of logical explanation, can now be given a satisfactory interpretation. To bring the subject to that conclusion, it will be necessary to search eagerly for all possible indications concerning the physiological bases of morphogenesis. Normal features must be comparatively considered, paragenesis and regulation must be elucidated, young and later stages must be functionally correlated. In a word, Unity in space and time must be our Ariadne's thread.

REFERENCES, TERMINOLOGY, ACKNOWLEDGMENTS

A detailed account of many researches is to be found in the above cited books and some others.[1] Only recent or specially significant investigations will be referred to at some length. The names of the authors will be generally given in paginal notes. The simple ones concern a quotation of the mentioned fact. Those preceded by "cf." indicate a recent publication which can be used by the reader for the related literature.

The terminology used makes a distinction between the groups of cells and layers still capable of further segregations and those which have reached the end of their embryonic evolution. The former will receive designations with the suffix -*blast*, the latter with the suffix -*derm*. In the case of Chordates, we shall have to consider, at the end of gastrulation, an *ectoblast* which later splits into neural plate (*neuroblast*) and *epiblast*; a *chordo-mesoblast*, soon separated into *chorda* and *mesoblast*, which itself gives the *somites*, the *nephrotomes*, the *coelomic linings* and *mesenchyme*; and an *entoblast*, matrix of the whole digestive and respiratory

[1] Cf. Dalcq, 1935 a.

tract. In the Amniotes, the outer layer of the didermic stage preceding the primitive streak will be designated as *primary ectoblast*. In Invertebrates, the terms *ectoblast*, *mesoblast* and *entoblast* will be used with their usual meaning.

The present work has been undertaken at the solicitation of Mr C. H. Waddington. His kind request has incited me to an inquiry which has led me much farther than I could foresee. No pleasure is more valued by a scientist than a better comprehension of the field which he is exploring. For having given me this high satisfaction, I heartily thank my distinguished colleague.

I am also deeply indebted to Dr G. Vandebroek for the loan of unpublished results and documents; to Dr E. Van Campenhout, Dr L. Van den Berghe and Dr J. Pasteels for reviewing the manuscript and helping with most valuable suggestions.

CHAPTER II

A brief outline of Organogenesis, especially concerning Chordates

It has long been recognized that the initial processes of animal development show **common fundamental characters.** Their uniformity is nearly complete in regard to meiosis and the growth of the egg. Maturation and fertilization already tend to be more specialized, although not very deeply. Then there always follows a segmentation through karyokineses and—without speaking of other changes inherent to the same period—the blastomeres thus acquire the mobility and laxity that make morphogenetic movements possible. The orientation and chronology of these movements vary according to the species. Everywhere, except in certain Sponges, this period results in the enveloping of cellular groups by an external layer. The uniformity of this phase induced Haeckel to imagine the ancestral type called *gastraea*. If such a being ever really existed, its physiology must have been radically different from that of our actual gastrulae. In fact, these do not constitute a larval stage. Except in the case of some Hydroides, where growth is often extremely precocious,[1] and in that of placentary Mammals, where the elaborated uterine secretion seems to be absorbed by the young blastomeres, the cells of the gastrula do not nourish themselves nor grow noticeably. They divide and they move, and the objective they pursue unceasingly is the construction of the primitive organs. These vary a good deal in the case of Chordates, Worms, Molluscs, Insects, etc., and it will be advisable for the moment to limit ourselves to the first-mentioned group. We shall add the necessary indications when we have to consider other types of ontogenesis.

The termination of gastrulation is, for all Chordates, marvellously uniform. It is the **constitution of a young embryo** provided

[1] Cf. Teissier, 1931.

with five or six very characteristic organs, each having its typical form and size (figs. 1, 3, 41). A *neural tube*, swollen at its anterior part into a cerebral vesicle, lies above the *notochord* and protrudes somewhat beyond this, except in the case of *Amphioxus*. These two organs stand over the elongated pocket of *entoblast*, which encloses an archenteric cavity somewhat constricted in its forepart, the pharyngeal pouch. The *mesoblast* is moulded on the sides of the neural tube, of the chorda and of the archenteron. A continuous *epiblast* surrounds the internal parts.

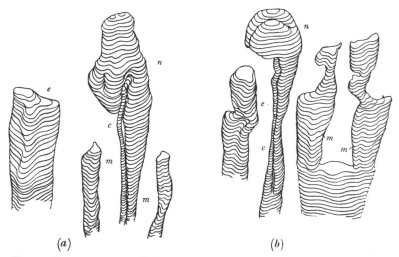

Fig. 1. Reconstruction of the main organs of rabbit (*a*) and tortoise (*b*) embryos, at a comparable stage. *c*, chorda; *e*, entoblast; *m*, mesoblast, mostly somitic; *n*, neural organ.

This stage, common to all Chordates without exception, is moreover of a short duration. Among Prochordates, the Ascidiae immediately show a scission of the mesoblast in *mesenchyme* and *myoblast*, containing the future muscular cells.[1] In *Amphioxus*, the same mesoblast remains partly continuous with the archenteric wall; it takes the aspect of the well-known saccules, at first symmetrically disposed; later, when these saccules have been pinched off, the perfect symmetry is modified by the appearance

[1] The "caudal mesenchyme" is simply a posterior group of less differentiated cells, later transformed into muscle cells.

of diverticles of the future pharynx, a larger right one displacing the right saccules slightly backwards. In the young embryos of Vertebrates, cephalic neural crests appear on each side of the brain, while the cerebral vesicle constricts in three distinct parts, the anterior one expanding laterally into primary optic vesicles. A loose mesoblast, mostly of mesenchymatous aspect, fills up all the free spaces of the now quite distinct head. In the trunk, the dorsal mesoblast (epimere) shows a caudally progressing segmentary constriction into somites, which later on become entirely free from one another. The intermediate part (mesomere), corresponding to some of the first somites, forms a small block of cells, dorsally indented by the somitic segmentation. This rudiment of the pronephros preserves its continuity with the ventral mesoblast (hypomere), which fills the space between the lateral wall of the archenteron and the epiblast. The material of the *lateral plates* is soon separated, by a new arrangement of its cells, into an external and an internal layer, the *somatopleure* and *splanchnopleure*, forming the *coelomic linings*. The entoblast is now distinctly divided into a pharynx, the lateral wall of which is pushed outside in two, three or four branchial pouches, and a broader, nearly tubular enteron. The primordial organs, when fully constituted, are remarkably similar throughout Chordates (figs. 1, 41).

This accomplishment of a uniform result is, however, attained by a development which is clearly subject to considerable **variation according to the zoological position** of the species.

Between Prochordates and Vertebrates, we first notice that, at the blastula stage, the cells of the former are placed in a cuboid or cylindrical epithelium, while those of the latter, especially in Anamniotes, are arranged in a pluristratified wall (fig. 2). Such a difference is not only quantitative but also indicates that, in the second case, cleavage has been more pronounced before the appearance of the formative movements. Another contrast, much more important, consists in the position occupied, relatively to the axis of the egg, by the materials which are to form the various organs. But it would be premature to discuss this matter before having studied the functional features of the presumptive territories in the whole Phylum.

Considering the Vertebrates, the quantity and the density of the vitelline reserves affect the development in a marked way. It causes cleavage to be partial in the case of all telolecithic eggs and considerably complicates their gastrulation. But it is well known that Sauropsides present embryonic aspects (fig. 5) which are extremely different from those of Teleosts or Selachians (fig. 3). The site of the invagination, the appearance of the first visible

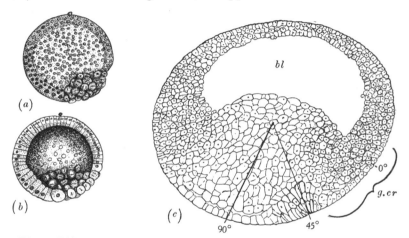

Fig. 2. Difference in the arrangement of the cells, at the incipient gastrula stage, between Prochordates and most of Vertebrates. (a) and (b) Very young gastrula of *Amphioxus*, whole mount seen from the left side (a) and in optic sagittal section (b). The conspicuous rounded cells, many of which are in division, are the mesoblastic ones. Redrawn from Conklin, 1932. (c) Early gastrula in *Discoglossus pictus*. The exactly sagittal section shows the place where invagination begins, exactly at equal distance from the equatorial plane and the vegetative pole. *g.cr*, average position of the gray crescent. *bl*, blastocoel.

structures, and the chronology of the processes are all affected, and an interpretation will only be possible by the judicious combination of the cytological and vital staining methods. It will, however, be convenient to examine separately the mode of **formation of the entoblast.** This question has been recently solved, in my opinion, by the study of serial sections. Its consideration will clear the way for a more general analysis.

In the case of Amphibians, it is perfectly evident that the entoblast is represented by the vegetative blastomeres. The blastoporal lip appears in the supero-dorsal region of this material, and the cells which are the first to invaginate will form the anterior wall of the pharynx (fig. 14). The rest of the material is gradually circumscribed by the lateral lips of the blastopore and builds up the roof, the walls and the floor of the archenteron.

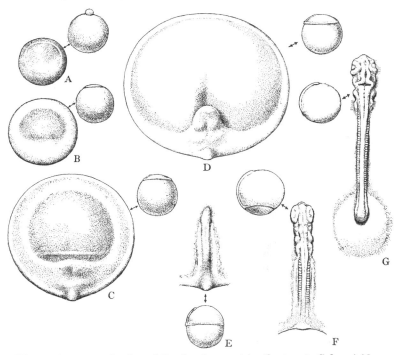

Fig. 3. A panoramic view of the development in the trout, *Salmo irideus.* For each stage, the illustration of the entire egg is accompanied by a drawing of the blastoderm or of the embryo, at a higher but constant magnification. A, Blastula stage. B, Very young gastrula. C, D, E, Invagination with progressive appearance of the embryonic mass. F, Invagination attaining the vegetative pole, embryo with terminal knot. G, Envelopment finished and blastopore closed, embryo with optic vesicles and optic pits. From Pasteels, 1936.

In germs where the yolk mass attains such a size and compactness as partially to inhibit cleavage, this accumulation of vitellus occurs in the most vegetative part of the egg, representing the main portion of the entoblastic area in holoblastic germs. The residual active entoblast thus takes the shape of a crescent placed on the edge of the blastodisc. In fact, gastrulation begins in a definite region of its periphery, and when the dorsal lip becomes visible, the migration of the thin entoblastic layer into the sub-germinal cavity is already well marked (fig. 4). The fact is positively established, both for Teleosts[1] and Selachians.[2] This

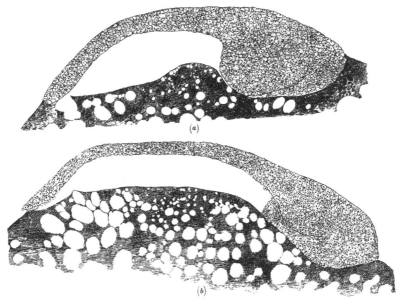

Fig. 4. Two sagittal sections of a trout blastodisc (*Salmo irideus*). (*b*) is 24 hours older than (*a*). In (*a*) invagination of the entoblast. In (*b*) the chordo-mesoblast continues the same movement, without being delimited from the entoblast. From Pasteels, 1936.

initial activity of the presumptive entoblast introduces the gradual formation of the embryo (fig. 3). In Reptiles, when cleavage of the blastodisc attains a certain stage, the peripheral region of the blastodisc, the *area opaca*, gradually spreads as a thin layer over the vitelline sphere, towards the equator (fig. 5). In the more transparent roof of the subgerminal cavity cells become condensed in an ill-defined zone, making possible the

[1] Cf. Pasteels, 1936.   [2] Cf. Vandebroek, 1936.

distinction between an *area pellucida* and an embryonic shield (fig. 6). At the posterior part of this shield a thicker condensation appears, the blastoporal plate (fig. 5). In its centre, one or more furrows are formed, at first ill-marked and variable, then clearly visible as a forward curved groove (figs. 5, 7). This is the real place of the entoblastic invagination: cells become elongated, infiltrate into the depth and migrate in a forward direction

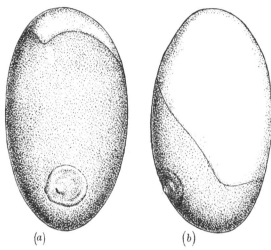

(a)   (b)

Fig. 5. Egg of *Lacerta viridis* at the beginning of gastrulation. Seen (a) from the animal region; (b) from the ventral region. The extension of the extra-embryonic ectoblast is already considerable; its limit is very distinct. The elevated embryonic region shows, as a clear margin, the *area opaca*; inside, the *area pellucida*, somewhat darker; and finally, in the centre, the *embryonic shield* (clearer), extended backwards by the *blastoporal plate*. The young, entoblastic *blastopore* is clearly visible.

(fig. 8). For a long time it was thought that entoblast was formed by a pure delamination at the roof of the subgerminal cavity. All species are certainly not equally favourable for the analysis of this process, but the observations made in tortoises, especially in *Clemmys leprosa*,[1] are extremely convincing. Before the appearance of the blastoporal plate, the embryonic territory is formed of a simple and continuous layer of cells, under which

[1] Pasteels, 1937.

some big yolk cells are dispersed in the subgerminal cavity. These few elements do not, or very rarely, divide, and they form at most only a small part of the entoblast. In the main, this layer takes its origin from the cellular flow which penetrates through the blastopore. These elements gradually spread under the surface

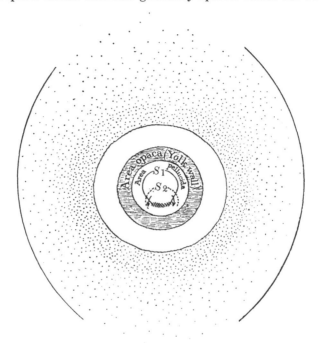

Fig. 6. Regions visible on a living Reptile or Bird egg at the beginning of development. Inside the *area pellucida*, two curved lines, S_1 and S_2, show the vague limit of the embryonic shield, in Reptiles and Birds respectively. The crossed lines indicate the blastoporal plate of Reptiles, sometimes visible in Birds as an *entoblastic area*. The blastodisc expands over the yolk in a thin layer (empty girdle) outside the *area opaca*, down to the enveloping border. The still naked yolk is stippled.

of the embryonic shield in a delicate single layer. Simultaneously, some posterior material of the blastoporal plate penetrates into the neighbouring region of the yolk wall, dorsally limiting the subgerminal cavity (fig. 8); they remain there quiescent during the whole period of gastrulation and finally transform into the

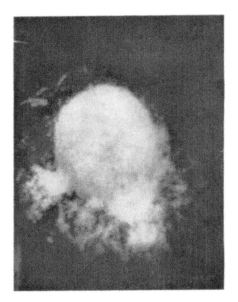

Fig. 7. Photograph of a blastodisc of *Clemmys leprosa* detached from the vitellus at the beginning of gastrulation. At the posterior margin of the disc, a whiter, elliptic, slightly protruding area is the blastoporal plate. It is depressed by the entoblastic blastopore, somewhat curved, with anterior concavity. Photograph of Pasteels.

to face p. 14

primordial germ cells.[1] Reserving any comment about this fact (cf. p. 45), we must note the intradiscal situation of the Reptilian entoblast, contrasting with its peridiscal position in telolecithic Anamniotes. The same conclusion is reached with Birds. In these, the proportions of the *area pellucida* and the embryonic shield are somewhat different, and the latter reaches the dorsal sector of the *area opaca* (fig. 6). There is no blastoporal plate. For a long time the entoblast was thought to be formed by a delamination. In the pigeon, it was suggested that an invagination occurs at the external border of the blastodisc, during the

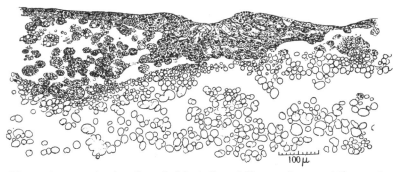

Fig. 8. A parasagittal section of a blastodisc of *Clemmys leprosa* at the onset of gastrulation, to show the invagination of entoblastic cells. Left, the subgerminal cavity, with already invaginated but dispersed cells. The entoblastic elements lying at the right of the blastoporal groove are the future primordial gonocytes. From Pasteels, 1937.

migration of the egg in the oviduct.[2] None of these assertions seems tenable.[3] The entoblast of the hen's egg has recently been shown to form by polyinvagination: numerous small grooves appear on the surface of the very young blastodisc during the first hours of incubation. They reveal the place where cells migrate in the depth (fig. 9) and secondarily join together in a continuous layer.[4] The process is more active in the dorsal half of the disc. It is similar to that of Reptiles, but scattered over a larger territory.

A fundamental difference must thus be admitted between

[1] Pasteels, 1937, p. 164.   [2] Patterson, 1909.
[3] Pasteels, 1937, pp. 394 *et seq.*
[4] Mehrbach, 1935; Pasteels, 1937, p. 399.

Anamniotes and Sauropsides from the very beginning of morphogenesis. In telolecithic eggs of the former, invagination takes place at the margin of the blastodisc. In the Sauropsid eggs, the entoblast takes its origin from cells either situated near the centre of the blastodisc or scattered in a dorsal but still intradiscal territory. It is obvious that the yolk accumulation does not afford a satisfactory explanation of this important modification, which has, as will appear later, other consequences (cf. p. 93). But would these considerations apply to Mammals?

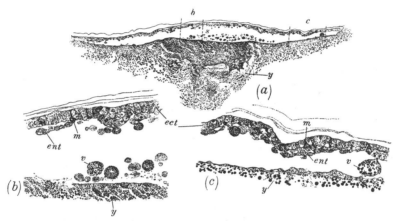

Fig. 9. Formation of the entoblast in the chick. Section of an unincubated blastoderm. (*a*) General view of the blastoderm, the subgerminal cavity and the yolk. (*b*), (*c*) Two regions of the blastoderm at a larger magnification. *ect*, primary ectoblast; *ent*, entoblastic cells already entered in the subgerminal cavity and preparing the entoblastic layer; *m*, wandering cell, charged with yolk, joining the entoblastic layer; *v*, large vitelline cells probably detached from the deeper part of the blastoderm during cleavage; *y*, yolk; *s*, subgerminal cavity. From a preparation of Pasteels.

It is certainly remarkable that an entirely similar entoblast formation has already been described in Marsupials. Just before morphogenesis, the blastocyst of *Dasyurus* is formed of a one-celled layer. Studied in whole mounts, after correct flattening, it shows some darker cells among the more usual clear ones. These special elements gradually sink into the blastocoele and build up the entoblastic layer of the didermic stage[1] (fig. 10). The

---

[1] J. P. Hill, 1910, p. 54.

formation by delamination is thus only maintained in the classical
descriptions for Mammals other than Marsupials. But even there

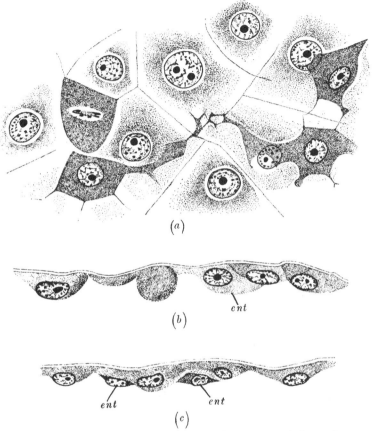

Fig. 10. Origin of the entoblast in young blastocysts of the native cat
(*Dasyurus viverrinus*). (*a*) A part of the embryonic area, flattened, with the
dark-coloured entoblastic cells. (*b*) Section of the same area when formed of
a uniform layer; *ent*, entoblastic cell. (*c*) Section after migration of the entoblastic cells under the primary ectoblast. Redrawn from J. P. Hill, 1910.

it may be remarked that the cytological data are not really
conclusive. Migrations of cells may probably have escaped
observers less aware than we are to-day of their frequency and

of their importance. The peripheral region of the embryonic area, where a close contact often exists with the trophoblastic layer, should especially be studied in this respect. In any case, the generalization of the migratory entoblast formation in Amniotes is not actually excluded.

This question of the entoblast formation being so far settled, it will also be useful to clear up another much disputed point, the part played by **mitotic cell divisions** in the morphogenetic process.

Growth being nearly everywhere bound to karyokinetic activity, there is a natural tendency to consider that the formation of the young embryo is also the consequence of more or less localized proliferations, analogous to those observed in a young limb bud. Indeed, karyokinetic divisions are the prominent feature of the period immediately following fertilization. When gastrulation begins, they certainly do not stop and are still found in all parts of the germ. Are they specially concentrated in some regions, so as to provoke changes of their form? Most decidedly not. Beside the negative impression obtained from slide study, precise counts have been made in two important and favourable cases. In the trout egg, the distribution of karyokineses has been studied in the late blastula and the gastrula without observing any noticeable difference in the regional frequency of mitotic division. Then, the region where the young primitive streak of a chick egg is growing has been compared with the lateral ectoblast. To avoid counting twice the same nuclei, each second section has been discarded. In the primitive streak, 75 divisions have been found among 3090 nuclei; in the primary, lateral ectoblast, 75 among 3069 nuclei. The equality is obvious.[1]

In the case of Amphibians, a recent theory has again attributed to cell division a rôle at the onset of morphogenesis, and it has been suggested that a "mitogenetic centre", real initiator of morphogenesis, exists on the dorsal part of the blastula.[2] This assumption is chiefly based on the relative smallness of the dorsal blastomeres. But this fact has quite a different meaning. During the first steps of cleavage, the mitotic apparatus has a

[1] Pasteels, 1937, p. 415.   [2] Wintrebert, 1933.

tendency to be slightly repelled dorsally, in such a way that the dorsal blastomeres are more or less reduced. For that reason, they acquire some rhythmical advantage on the ventral ones. Cellularization thus happens faster in the dorsal part of the egg, where the cellular movements will appear first. This coincidence is certainly connected with the organization of the egg, but nothing points to the existence of a mitogenetic impulse. At the beginning of gastrulation, local proliferations are not to be observed, except in special cases, as the *Amphioxus* mesoblast. Cell divisions exist, of course, but they only accompany, without having any apparent effect on, the active cellular displacements which are the dominant feature of gastrulation, especially in Chordates. Later, when the main organs are individualized, local centres of proliferation will indeed appear, and play an important rôle in the definitive conformation of the organs. Certain of those centres, characterized by their abundant mitotic divisions, will be only temporary, others will persist and even increase in the young animal. They will finally be integrated in the growth pattern of the individual. How these secondary processes of growth are related to the primary events of organogenesis, which are the actual objects of this study, is not yet clear. Local advantages of nutrition or respiration resulting from peculiar dispositions fortuitously realized in the three-layered embryo may have some epigenetic influence. But causes more intimately bound to the constitution of certain cell groups are probably more efficient. It will become easier to understand the origin of the general and local growth fields[1] when the morphogenetic organization of the egg has lost most of its enigmatic character.

[1] Cf. Huxley, 1935.

CHAPTER III

The kinematic aspect of embryogenic processes in Chordates

It is certainly erroneous to consider sciences as constructions methodically erected according to a design thought out at leisure. Like all human achievements, they are a fruit of life, and of progress in those directions which happen, at a certain moment, to be favourable. The connection between the discoveries thus made is then, in many cases, a late acquisition. It has been asserted[1] that Embryology should attain the limit of observation before entering the experimental path. But as a matter of fact, the experimenter opens the eyes of the observer. In the typical instance of Amphibians, it was the problems resulting from various experiments which made Vogt conscious of the pressing necessity of exactly determining the position occupied in the egg by the materials which form the various organs. This was performed by using the **vital staining method.** The few cells laden with Nile blue, or neutral red, or Bismarck brown, retain their entire activity and they can be traced step by step from the germ into the embryo. It is easy to note the direction and speed of their displacement and to conclude, from their final location, the presumptive value of the spot on which the coloured mark was placed. The representation of development thus arrived at is evidently far superior to the ancient description of successive stages, more or less arbitrarily selected. These were only temporary aspects, which were linked together as seemed most probable to each observer. The fine researches of Vogt have brought the description of Amphibian development to a standard precision. Immediately completed by accurate investigations on other Chordates, they have helped us out of the meshes of the two-layered gastrula, of the generation of the third layer from

[1] Wintrebert, 1935, p. 9 and *passim*.

the inner one, of the closing of the blastopore by concrescence of its lips, of the formation of trunk and tail by appositional growth, etc. These new, difficult, but extremely profitable investigations have rendered apparent, firstly in the development of Amphibians, then in that of other Chordates, a unity, a harmonious submission of parts to the whole, that was absolutely unknown. They have revealed the laws of topographical distribution, and of the dynamic evolution by which the germinal territories are wholly controlled.

That early development of Chordates would obey common fundamental laws was evidently to be predicted as a corollary to the evolutionist theory. But it seems strange that, in spite of their devotion to that great Haeckelian idea, embryologists have been so long in discovering the true Ariadne's thread between the various aspects of the first ontogenetic stages, inter-comparison of Prochordates, Fishes, Amphibians, Reptiles, Birds and Mammals. The reasons for such delay were: the imperfection of a study which was too static, too limited to conventional stages; then, for the recent period following Vogt's fundamental publications, the extreme difficulty of precise observation on certain eggs, e.g. those of Birds; and, finally, the erroneous choice of the principles used in detecting homologies. The situation remained rather perplexing while the well-elucidated case of Amphibians was only supplemented by that of the chick egg, which both Gräper and Wetzel had endeavoured to analyse from a really dynamic point of view. But they were actually attacking one of the last terms of a complex series, and encountered difficulties which they were unable to solve entirely.

The problem had to be faced at its basis. The study of *Petromyzon*, by Weissenberg, was a first really valuable step. It induced me to think that the second step ought to be an investigation of the eggs laden with much yolk and showing partial cleavage. It was a real joy for me to see this project thoroughly carried out by Pasteels, who discovered the map of the anlagen and the orientation of the morphogenetic movements successively in Teleosts, Reptiles and Birds. About the same time, Selachians were successfully studied by Vandebroek, and *Fundulus* by Oppenheimer. Adding that, recently, some supposed discrepancies concerning *Discoglossus pictus* have also been elucidated by

Pasteels, that Tung and especially Vandebroek have succeeded in using the vital staining method in the case of Ascidians, we have at hand the necessary elements for discovering what could perhaps be styled the morphological laws of early organogeny in Chordates. These should make more intelligible the general and local modifications by which an already segmented egg builds up an embryo possessing, in the right place, all its characteristic organs, ready for the cytodifferentiation and consequently for functional activities.

Classical Embryology generally divides this continuous transformation into two distinct events, **gastrulation and neurulation.** The first term draws attention to the blastopore, the second to the neural plate. Conspicuous as these morphological features may be, they do not afford a firm basis for the distinction. From a morphological viewpoint, the blastopore, *sensu stricto*, mouth of a primitive digestive cavity, is not exactly constant. From a physiological viewpoint, the phenomena influence each other continuously without any fundamental change in their causality. The terms gastrulation and neurulation are of course not to be discarded, but they are only to be used for the sake of convenience. It seems more advisable, at least for didactic purposes, to consider a primordial organogenesis, forming the five or six elementary organs of any Chordate, and a second phase, causing more localized and specialized complications. In fact, we are bound to do this, because these late transformations are different in Prochordates and in Vertebrates. In this work, Prochordates can only interest us for their primary organogenesis. Among Vertebrates, the second phase also develops along uniform lines. In the entoblast, we observe the general elongation leading to the formation of the digestive tract and the moulding of the branchial pouches; in the chordo-mesoblast, the isolation of chorda and somites, of the curiously limited pronephritic rudiments, and of the coelomic layers with the incipient limb buds; in the ectoblast, the formation of the brain vesicles, sense organs, neural crest and placodes. But except for the degree of complication, no fundamental distinction is to be drawn between primordial and secondary organogenesis. This unitary conception will find its justification as much in the accurate study of normal ontogenesis as in the experimental analysis of its causality.[1]

The germ is a unit, in space and in time. What we are interested in is its formative activity, its fascinating struggle in escaping from the sphericity or similar state imposed by banal physico-chemical conditions and in building up an organized embryo that will be soon capable of dominating its environment. What we have to study is the whole deformation, by its own strength, of an initially rounded system; and cleavage itself cannot be separated, except arbitrarily, from morphogenetic processes. It already affords conditions indispensable to formative activities. Leaving aside the morphogenetic effects of the nucleus and other functional changes to be discussed later, the reduction of cell size is in itself important. A first displacement of some materials as a mechanical consequence of the cleavage must also be noted. In Anurans, it has been shown[2] that the gradual expansion of the blastocoele provokes an appreciable descent of the dorsal presumptive territories.

The true **morphogenetic movements** are essentially active. Cell groups endowed with a definite presumptive value migrate in a well-ordered troop toward their precise aim. They act as would amoebae forming their pseudopods constantly on the same side, and progressing all together in one direction. But no true pseudopod is actually formed in the case of embryonic cells, their bodies change their shape as a whole and gradually move with regard to contiguous and less active or otherwise orientated groups. A certain autonomy has been found in these tactic operations, but it would be an exaggeration to consider the progressive change of shape in the germ as the mere summation of local transformations. Its own general formative activity, overriding the host of local tendencies, must not be lost sight of.

Vogt has described in Amphibians five types of specialized

---

[1] This point of view is not inconsistent with the gradual and considerable modification of the metabolism during development. In the case of Amphibian egg, it has been demonstrated that the neurula stage becomes distinctly more sensitive to deprivation of O_2 and to the monoiodacetic acid (J. Brachet, 1934).

[2] Votquenne, 1933, 1934; Vintemberger, 1933.

movements, bound together by a sort of internal logic. The most characteristic, because intimately connected with inductive influences, is *invagination*, i.e. invasion of the blastocoele by the lower materials of the egg. Entoblast, as we have seen above, is at first largely involved, but soon its movement is followed by an intense tucking of the chordal and mesoblastic materials round the lip of the blastopore. Invagination could not possess any meaning if the rest of the blastula did not undergo the extension necessary for enveloping the material rolled inside: such is the rôle of *epiboly* in the ectoblast, which is precisely the territory capable of superficial extension but lacking any faculty of invagination. The combination of these two first movements would be sufficient to close the blastopore, but the germ nevertheless would remain spherical. The dorsally localized activities are thus essential to morphogenesis. The first one is a *convergence* of the whole dorsal material, entoblast, chordo-mesoblast and ectoblast, towards the medio-dorsal line. It should result in a heaping up and a dorsal bulging if it were not counteracted by a powerful *extension*, which lengthens the embryo and makes its cephalo-caudal axis apparent. Finally, a ventral *divergence* is also to be recognized; certain parts of the mesoblast, which are the last to pass through the blastopore in a convergent movement, are destined to diverge secondarily toward a lateral position. Not very pronounced in Amphibians, this movement acquires a larger importance in higher animals. The prevalent features are, however, epiboly and invagination, convergence and extension. Of these, the two first affect definite territories, the two last mostly concern the whole dorsal region of the germ. Deformations of the entire gastrula are not especially conspicuous in Amphibians, but they are quite evident in Birds, as will soon appear.[1]

Leaving aside the still unexplored case of Mammals,[2] let us summarize gastrulation in Prochordates and Anamniotes and the homologous episodes of development in other Vertebrates. Actual observations now allow us to assert that everywhere these same four or five types of morphogenetic movements govern early development. Of course, differences, as much of

[1] Pasteels, 1937, p. 412.
[2] The recent improvement of culture methods by Jolly and Ferester-Tadié, 1936, will soon allow the necessary researches in this field.

topographical as of chronological order, and mostly of the latter kind, are to be noticed. The most frequent is the precocious migration of the entoblast, which has already been described. In some cases, especially with Birds, the extension occurs particularly late. Other details will be given when examining comparatively the various maps. But, before that, we must become acquainted with the changes of shape that occur during neurulation and the consecutive stages.

We may consider successively the three layers already individualized at the end of gastrulation. In the *entoblast*, we simply observe a general elongation and, anteriorly, a more active growth of the inner surface of the lateral walls, resulting in the branchial folding. Analogous but more localized processes, where cell proliferations play their part, occur later, on the ventral aspect of the pharynx, to form the thyroid, the respiratory apparatus; digestive glands also result from a local budding, but the kinetics of these late complications are not studied. In the middle layer, the transformations obey two distinct tendencies. In the *chorda*, we simply see a lengthening due to rearrangement of the cells and their gradual vacuolization, which represents the first cytodifferentiation. In the *mesoblast*, cells show a marked growth of their surfaces, but with differences that considerably influence the shape realized at various levels of the dorso-ventral axis. Dorsally, the splitting into somites is observed, the first formed of which may be practically considered as the boundary between head and trunk in the Vertebrate embryo. It looks as if, when a certain number of cells have attained some kind of maturity, they become entangled in some substance of high surface tension and tend to round up as a small mass. An identical process would be successively repeated for more posterior groups. The same mechanism seems also to be effective, but to a lesser degree, for more lateral parts. Notches indent the mesoblastic layer to limit the stalk of the somites, i.e. the small nephrotomes. On the other hand, the tendency to free surface growth is especially expressed by the invasion of all remaining spaces by actively amoeboid cells, the mesenchyme. This process is of considerable importance in the head; and when building the three vesicles that are the matrix of the eye muscles, it recalls the formation of somites. But mesenchyme also appears, by active migration, between the chorda, the entoblast and the somites, then on the whole surface of these, and also on the deep and superficial aspects of the two coelomic linings: its cells form a filling material apparently quite homogeneous, in spite of its future considerable diversification. In the next period, the somitic cells, excepting the superficial ones specialized in the formation of the aponeuroses, lengthen in the cephalo-caudal direction, show a multiplication of their nuclei and attain their fibrillar differentiation. The nephrotomes corresponding to certain anterior somites—with considerable variation according to the groups—especially augment their exterior surface, in such manner that each one elongates into a pronephric canalicule, which fuses with the adjoining one to form the segmental duct or primary ureter. This canal seems to incorporate most of the material of the more posterior nephrotomes, excepting what is to become the mesonephros. About the same time, the mesoblast of the mesobranchial field forms a coelomic cavity and some cells of the splanchnopleure build up the rudiment of the cardiac tube, where rhythmic contraction immediately begins, while fibrillar differentiation appears. Heart, gills, pronephros and myotomes almost simultaneously initiate their combined activity; the vital importance of this fact does not need to be emphasized. In the *external layer*, the segregation of the neural organ (neuroblast) and epiblast out of the ectoblast also results from the differential growth of certain free surfaces. In the whole territory overlying the archenteric roof, i.e. chorda and somites, the cells of the so-called sensory layer, the deepest ones of the still thick ectoblast, seem actively concentrating, and an actual convergent movement of more lateral elements has been recently demonstrated.[1]

When in place, the neuroblast cells elongate in small pyramids standing with their base on the underlying chordo-mesoblast. From these individual changes, the folding of the neural plate results; then, by a continuation of the same process, it closes into a tube, only leaving open anteriorly a neuropore and posteriorly a neurenteric canal, which is absent in some cases. This development of the neural organ has been recently[2] attributed to a contraction. It seems more advisable to use this word only for

---

[1] Dettlaff, 1936, p. 203.
[2] Wintrebert, 1931; 1935, p. 65.

muscular activity, where a cell, considerably lengthened thanks to its internal structure, suddenly comes back toward sphericity, by a kind of rebound. In the recently fertilized egg of many species, a similar process is indeed observed, which foreshadows the future contractile specialization. But, in the formation and closing of the neural tube, no such instantaneous and reversible change is involved. There is simply a change of shape related to a definite deformation of the cells.

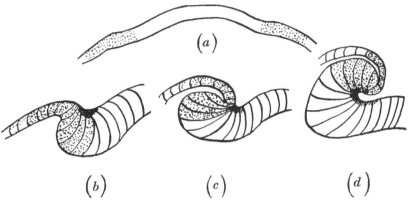

Fig. 11. Origin of the neural crest cells in the Axolotl. Schematic representation of four transverse sections of the ectoblast, with indication (stippled) of the cells that will later migrate to form the neural crest. Redrawn from Raven, 1931.

The cerebral vesicle is not yet closed when the neural crests appear. They are derived to a certain extent from the edge of the neural tube, and, for the rest, from the sensory layer of the epiblast (fig. 11). Like the true mesenchyme, they tend to increase their free surfaces and consequently become disseminated. It is a remarkable fact that mesoblast and ectoblast behave in a similar way, firstly forming main pieces, the somites and the neural plate, then giving off loose cells which will settle in the interstices left in the body. Finally, we have to mention the placodes related to the hypophysis, to the olfactory, optic and otic organs, to the head ganglia and the lateral line organs, all produced by the inside folding of the epiblast or the expansion of its sensory layer, i.e. again the growth of some localized deep surfaces.

This general and necessarily schematic survey of the dynamic aspects of development, during the prefunctional period, shows that the changes are not due to local increase of volume or proliferation, but to the **modification of shape** of the involved cells. Except for some progressive imbibition, the volume of these cells does not change; their shape only is modified. This means that, at the basis of each local complication, a process may be found that affects the surface of the cell and implies the growth of the surface film. Whatever the internal transformation of the physical and chemical situation may be in the cell cytoplasm, each step of development may be thought of as a growth of the cell surface, which I shall briefly refer to as the *auxetic surface process*. Its intensity varies a good deal, attaining its maximum in the expanded cells of mesenchyme. It is often differential, favouring sometimes the outside face, as in epiboly, or the inside membrane, as in the repeated invaginations, or the lateral walls, as in extension. The essential result is always the production of new surfaces where functional exchanges will readily take place.

CHAPTER IV

A comparative study of the presumptive territories in Chordates

In a normal egg, the materials which are to develop into the various organs have a definite location. If their shape, their volume and their exact situation were known, the germ could be ideally separated into a set of pieces fitting into each other. This remark does not imply any assumption concerning the particular properties of the lumps in such a mosaic. From this purely **topographical viewpoint**, causality is momentarily forgotten. In the case of Vertebrates, the term presumptive material is not absolutely exact, as it does not take into account the existence of an inductive substance to be transferred from the chordomesoblast into the ectoblast. Only the visible and especially the colourable elements are concerned. The maximum of precision in their study is obtained at the end of cleavage, when well-defined territories may be selectively stained. Evidence concerning more or less deeply located materials is always scarce, but can be obtained by certain expedients. We intend to consider comparatively the data actually available for the various Chordate eggs just before gastrulation. A complete study of the matter is, however, not necessary. Any specialized worker will indeed refer to the original papers, and we only wish to show, through the comparison of the maps, the unity of general design and the seriation of the aspects observed in Chordates.

By reason of their average type of development, and of the complete data available for them, the small-egged **Amphibians** give the norm of our knowledge concerning presumptive territories. The perfect maps produced by Vogt in the case of *Pleurodeles* and *Bombinator* (fig. 24, *b*) have too often been described to be analysed here. We shall choose the egg of *Discoglossus pictus*, which will give us a basis for the appreciation of experiments performed on it. The map (fig. 12) shows the two characteristics of Anurans: a relatively high situation of the marginal zone, and a real narrowness of the neural field, which is far from reaching the animal pole. The indicated mesoblast is mainly somitic; the material of the coelomic linings and that of the pronephros lie under the inferior boundary of the mesoblast, in close contact with the entoblast. The ectoblastic territory

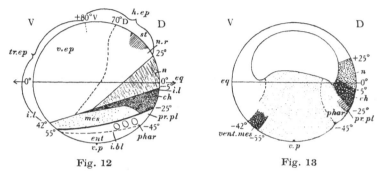

Fig. 12   Fig. 13

Fig. 12. Map of the presumptive territories on the surface of the *Discoglossus* egg. Left side view. *ch*, notochord; *ent*, entoblast; *h.ep*, head epiblast; *tr.ep*, trunk epiblast; *v.ep*, ventral epiblast; *i.l*, limit of invagination; *mes*, mesoblast; *n*, neuroblast; *n.r*, neural ridge; *phar*, pharynx; *pr.pl*, prechordal plate; *st*, stomodoeum; *i.bl*, virtual blastopore; *v.p*, vegetative pole. The broken line beginning at $+70°$ D is the lateral limit of the neural ridge: that at $+80°$ V points to the limit between head and trunk epiblast. From Pasteels, 1936, modified.

Fig. 13. Sagittal section of a young *Discoglossus* gastrula with limits of the anlagen in the depth. Same abbreviations as in fig. 12. *vent.mes*, ventral mesoblast. The two broken lines descending from the blastocoele floor indicate the average boundary between the big vegetative blastomeres and the smaller ones of the marginal zone. These were formerly higher and have descended during cleavage. From Pasteels, 1936.

certainly amounts to three-quarters of the surface. Its true neural part is relatively small, disposed in a crescent just above the chorda. The next part of the ectoblast is destined to form the neural ridge which surrounds, as a large flat band, the neural plate. The ventral limit between head and trunk epiblast is at $10°$ ventrally from the animal pole. The extension of these areas in the depth is illustrated by a sagittal section (fig. 13). As is also shown in fig. 2 *c* the dorsal lip appears at $-45°$, the ventral one at $-55°$, so that the line of the virtual blastopore is slightly inclined to the axis (fig. 12). The material situated just above the dorsal lip will form the pharynx. If it has been distinctly coloured, the superficial stained cells are found scattered among uncoloured elements (fig. 14), which indicates a considerable intermingling of cortical and deep material. This territory is thus responsible for the formation of the anterior part of the pharynx, of its floor down to the hepatic recess, and of its lateral walls with the branchial pouches. Above the pharynx anlage, between $-30°$ and $-25°$, is the narrow band of prechordal ento-mesoblast,

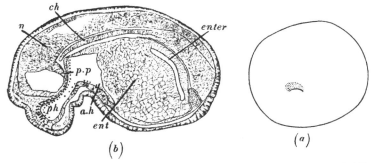

Fig. 14. A vital staining experiment on the *Discoglossus* egg. (*a*) A Nile blue mark colours the young dorsal lip. (*b*) Sagittal section of the embryo obtained. *a.h*, hepatic recess; *ch*, chorda; *ent*, entoblast; *enter*, enteron; *n*, neural tube; *ph*, pharynx; *p.p*, prechordal plate. The coloration is heavy and uniform in the prechordal plate, more spotty in the anterior wall and floor of the pharynx. From Pasteels, 1936.

extending obliquely towards the dorsal edge of the blastocoele. The chordal area extends from $-25°$ up to $-5°$, and the same material reaches the segmentation cavity. In the depth, the oblique limit between the notochordal and the neural areas crosses the equator.

With these data, the events of gastrulation can easily be followed. The marginal zone, which is the material to be tucked round the blastoporal lips, has appreciably migrated downward, as mentioned above (p. 23), during late cleavage. It forms, at the moment when invagination begins, a kind of belt broader on the dorsal side ($-5°$ down to $-45°$) than on the ventral one $-42°$ down to $-55°$). Its inferior limit will be called the virtual blastopore. The first movement to be noticed is an upward migration of the material which dorsally forms the floor of the blastocoele; it affects the surface at the blastoporal furrow, where cells are pressed together and elongated (fig. 2 c). The pharyngeal material is the first to be invaginated, and the still young blastoporal lip is soon formed by the prechordal plate flanked by mesoblast. At that moment, the four characteristic movements (cf. p. 24) enter in their phase of high activity. While chordal elements are progressively tucked in, they concentrate near the median line and pass in a narrow but long band into the roof of the archenteron. In the meantime, the movements have been extended to lateral parts. They are more easily accomplished in the small-celled mesoblast than in the heavy entoblast situated just above the virtual blastopore. While these big cells slowly move upwards, the mesoblast follows its own way in close contact with ectoblast. Hypomere material first crosses the blastopore and migrates in a ventro-animal direction. It is followed by the mesomere cells and the lower part of the somites which will become ventral, and finally by the upper dorsal part of the mesoblast. These last cells always preserve their original connection with the chorda. The displacement, at that very level, of the chordo-mesoblastic boundary, is most curious, showing signs both of invagination, convergence and antero-posterior extension.

At the moment now reached, the blastoporal furrow is comparable to a sickle and soon to a horse-shoe. The white, large plug of big-celled entoblast is in direct continuity with the already invaginated part of the same material. This forms the floor, the lateral walls and the anterior cul-de-sac of an archenteron which is still open caudally at the blastopore. The roof of the same cavity is built by the chorda and the somitic mesoblast. This last layer extends laterally between entoblast and ectoblast, but it does not reach the cranial extremity, so that a mesoblast-free field exists under the future pharynx. At this level, some of the most anterior mesoblastic cells will soon form the heart. During the same transformations, the ectoblast has not only been stretched over the internal material, but its epiboly has been combined with powerful convergence and extension. The neural area, originally disposed in a crescent, becomes concentrated and elongated in the dorsal region. Except for the lack of invagination, its evolution is quite similar to that of the dorsal half of the marginal zone. This harmony is known, as we shall see later, to be directly caused by induction.

The last steps of gastrulation are nothing but the continuation and completion of these correlated movements. Invagination proceeds up to its limit, the intucked material being now the posterior and even caudal mesoblast. The yolk plug disappears into the depths and the blastopore is closed, except for the persistence[1] of the anal orifice. The double-layered roof of the archenteron accentuates its convergence and specially its cephalo-caudal extension, which now applies to the whole embryo. The two lateral ridges of the entoblast extend toward the median dorsal line, under the somites and chorda, and fuse together. From that moment, gastrulation is practically at its end. The previously mentioned accumulation of cells in the external layer causes the neural plate to appear. It is separated from mere epiblast by a variable neural ridge, which includes the material of the neural crests, and, in the head, will undergo secondary changes to build up the facial epiblast.[2] At the posterior part of the neural plate, a small territory, proceeding from the lateral region where neural and chordal crescents join to the mesoblast (fig. 12), has still the dynamic tendency of the marginal zone and makes a late invagination. It completes the mesoblast of the caudal bud.

In **other Anamniotes,** holoblastic and meroblastic eggs must be successively considered. The only case of the first group which has been studied, the lamprey, shows many interesting features, which renders its map (fig. 15) very different at first sight from that of the Amphibians. The position of the anlagen on the dorsal line is, however, not changed relative to their neighbours. Epiblast, neural field, chordo-mesoblast and entoblast are regularly superposed. But the entoblast and epiblast are considerably augmented at the expense of the other areas. As a consequence

---

[1] In some species, the closure is complete and the anal orifice appears later by a secondary perforation.
[2] Halter, 1937.

of the dorsal concentration of the neural and chordo-mesoblastic areas, the epiblast is brought into contact with the entoblast. The mesoblastic girdle of the Amphibians is thus broken. The well-known aspects of the gastrulation and formation of the embryo in the lamprey are in agreement with that particular map. The typical formative movements take place with a considerable intensity in the dorsal region of the germ. Pharyngeal entoblast, prechordal plate, chorda, somites and lateral plate

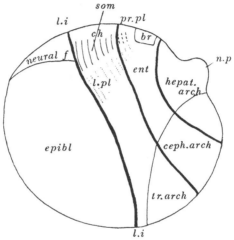

Fig. 15. Map of the anlagen in the egg of *Lampetra fluviatilis*, before gastrulation (nose stage). *br*, branchial area; *ceph.arch*, part of the entoblast destined to the head; *ch*, notochord; *ent*, entoblast; *epibl*, epiblast; *hepat.arch*, part of the entoblast forming the liver; *l.i*, limit of the invagination; *l.pl*, lateral plate (hypomere); *neural f*, neural area; *n.p*, nose process, a local deformation of the entoblast; *pr.pl*, prechordal plate; *som*, somites; *tr.arch*, part of the entoblast going into the trunk. The heavy lines separate the three main parts of the entoblastic mass. Redrawn from Weissenberg, 1936.

are progressively tucked in in the usual order. Convergence is indeed reduced, but cephalo-caudal extension is remarkably intense, so that the axial organs describe a curve round the entoblastic mass. Epiboly is also combined with dorsal extension. But a ventral lip does not appear, and no intucking happens there, because mesoblast is entirely absent.

Among meroblastic Fishes, most accurate observations have been made on the egg of the trout. Being relatively flattened in

young stages, it is a favourable material for the representation
of the movements (fig. 16). The virtual blastopore is the actual
limit of the blastodisc. Formative movements begin in the
dorsal region by the tucking in of the prechordal plate, and soon

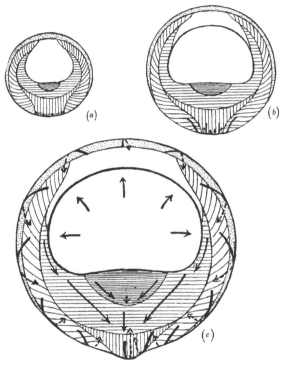

Fig. 16. Topography of the presumptive areas of *Salmo irideus*, with indication of the movements. The plain arrows indicate the movements in the superficial materials, the interrupted ones, in the already invaginated elements.
Wide vertical shading: notochord; close vertical streaking: prechordal plate;
wide horizontal shading: trunk neuroblast; close horizontal shading: brain;
oblique lines: somites; stippled: lateral and ventral mesoblast. From
Pasteels, 1936.

of the chorda, with the first somites. Progressively, lateral and
ventral parts of the virtual blastopore become real and the
vitellus is enveloped (fig. 3). The order of the invagination, the
direction of the movements, and their consequences for the

successive positions of the areas, are not essentially different from what has been studied in Amphibians, except for the limitation of the archenteron. The topography is also strikingly similar for ectoblast and mesoblast. The shape of the neural and chordal crescents, and also that of the mesoblastic girdle, are simply changed according to the flattening of the mapped surface. Comparison with Selachians (fig. 17) confirms the general significance of these observations. There is no difference but the

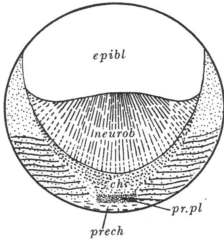

Fig. 17. Topography of the presumptive areas in *Scyllium canicula* just at the beginning of the invagination. *ch*, chorda; *prech*, prechordal entoblast; *epibl*, epiblast; *neurob*, neural area; *pr.pl*, prechordal plate. From Vandebroek, 1936.

extent of the neural area. We could also consider the plan detected by Oppenheimer in *Fundulus* by an investigation of cell lineage based on vital staining. In spite of the difficulties presented by this small-sized material, the agreement is quite satisfactory.

Nothing is easier to imagine than the transformation of these last maps into those of Amphibians. The yolk mass must simply be reduced and what remains incorporated into the embryonic cells, especially into the entoblast. The change of curvature will cause the desired modification. But, if Cyclostomes are considered, at least the lamprey, the variation of yolk alone does not suffice to bring about a resemblance of the maps. The difference is here deeper, the marginal zone being concentrated on the dorsal surface. The influence of a second factor on the relative position of the presumptive areas is pertinent. This is not, as will soon appear, a remark deprived of importance (cf. p. 93).

Before considering the Sauropsides, let us examine the case of **Prochordates,** especially that of simple Ascidians. Who would have imagined, when E. G. Conklin described the famous pigmented crescents of *Styela* and *Ciona*, that Nature then revealed

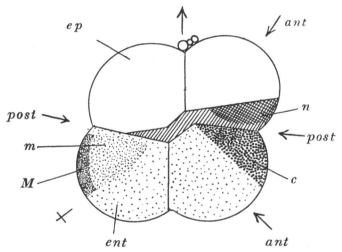

Fig. 8. Topography of the organ-forming areas in the 8-cell stage of *Ascidiella scabra. c*, chorda; *ent*, entoblast; *ep*, epiblast; *M*, myoblastic area, and *m*, mesenchyme area of the yellow crescent; *n*, neural area, the brain region cross-hatched. *ant* and *post* refer to the location of the materials after gastrulation. Original drawing of Vandebroek.

to him the essential features of the germinal organization in Chordates? The organ-forming territories that the great biologist recognized, thanks to the naturally pigmented areas, have recently been investigated, in *Ascidiella*, within the aid of local vital staining,[1] a research confirming and making precise Conklin's

[1] Vandebroek, 1937 (unpubl.).

discoveries. The exclusive use of that method implies, no doubt, serious difficulties for such a delicate egg. It must be directly stained, without the protection of the thin chorial membrane, left in place when Vertebrate eggs are concerned.[1] The marks must be put on a still undivided or slightly cleaved egg. But the stain is absorbed by yolk and other granules, and its stability is quite sufficient. The natural yellow coloration of the muscular and mesenchymatous material helps a great deal. At the 8-cell stage, the specialized regions are located at the level of the latitudinal furrow (fig. 17). On one side, which will be named dorsal for homology with Amphibians, the chordal area has the shape of a crescent, just at the upper part of the two dorsal macromeres. It is much shorter than the neural field, also crescent-shaped, which lies in the lowest region of the micromeres. On the ventral side, the broader, yellow-pigmented mesoblastic crescent occupies an important and rather thick part of the macromeres; the medial zone forms the myoblasts, while the horns form the mesenchyme. The rest of the egg is distributed between entoblast and epiblast. The thickness of the organ-forming materials can only be approximately stated. We may note, however, that the neurochordal material—conspicuous in *Styela*, from the 2-cell stage, through the presence of a light grey pigment—is just a thin layer covering entoblast. The mesoblastic material, crowded with the mitochondrial granules of the yellow pigment, constitutes, on the contrary, a really thick mass. This topographical situation of the organ-forming materials— using that word without any causal connotation—at the 8-cell stage is not appreciably modified by the continuation of cleavage. Gastrulation is heralded by a general flattening, which already involves some epiboly and invagination. A dorsal lip soon becomes visible. While the blastopore is gradually closed, the displacement and the change of shape of the marks reveal the convergence and the antero-posterior extension, with some interesting particularities.[2] No comparable study has been made, with vital staining, on *Amphioxus* eggs. But the cytological and experimental data gathered by Conklin authorize the assertion

[1] Except with Teleosteans and Selachians, where the marks are also placed on the naked eggs.
[2] Vandebroek, 1937 (unpubl.).

that its organization is quite similar to that of Ascidian germs. The precocious invagination of the mesoblastic elements (fig. 2, a, b) is perhaps to be noticed as a token of an early specialized activity due to the presence of the mesoplasm.[1]

As will appear later (p. 126), this particular plasm is indeed the element which seems to draw a sharp line between Prochordates and Vertebrates. Otherwise, the position of the anlagen along the sagittal dorsal line is perfectly identical and explains, with the occurrence of the typical formative movements, the resemblance of the larval stages. Reserving the meaning of the mesoplasm, we are thus allowed to speak of a unitary germinal design in Prochordates and Anamniotes.

With the Amniote eggs, especially in the **Reptiles,** we again encounter telolecithic germs, but their gastrulation is no longer marginal. Let us consider again (p. 13) *Clemmys leprosa,* which

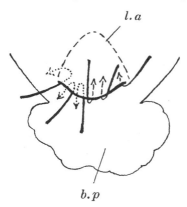

Fig. 19. Orientation of formative movements in the blastoporal region of *Clemmys leprosa.* The interrupted parts of the arrows indicate the displacement effected in the depth. *b.p,* blastoporal plate; *l.a,* limit of the anteriorly invaginated material. From Pasteels.

shows in the blastoporal region the most unequivocal phenomena. After the invagination of the entoblast (p. 14) all surrounding cells converge actively towards the blastoporal lip (fig. 19). Penetrating inside, they form, anteriorly to the blastopore, and as a direct prolongation of it, a chordo-mesoblastic canal, formerly

[1] Changes subsequent to fertilization will be considered p. 108.

known as the head process. Its walls are at first thick, and its floor is intimately superposed on the thin entoblastic layer. But the lower cells soon migrate towards both sides, so that the canal communicates freely with the subgerminal cavity (fig. 20). The structure now obtained is similar to the stage of Amphibian gastrulation, where the entoblast is still open dorsally by a broad

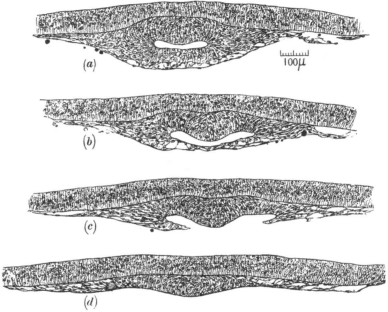

Fig. 20. Four transversal sections of the blastodisc in *Clemmys leprosa*, showing from the posterior (a) to the anterior (d) level the opening of the canal through a ventral cleft. Note the spaces appearing between the cells in (a) and (b). The entoblast is a very thin, unicellular layer. From Pasteels, 1937.

cleft and the archenteric roof only formed of chorda and meso-blast. In both groups, such a temporary disposition disappears by a gradual fusion of the sharp entoblastic edges. The archenteron being now dorsally closed, chorda and somites acquire their individuality, while inducing the neural plate out of the overlying ectoblast. Local vital staining being specially hard in the case of Reptiles, their map is still only approximate. But, as such, it shows again (fig. 21) the typical pattern of the dorsal region: the chorda inserted between the anteriorly placed neural area, the posterior entoblastic ellipse and the mesoblastic wings on both sides. This general resemblance with the Anamniote maps does not, however, explain the surrounding of these main areas by a continuous girdle of epiblast (cf. p. 93).

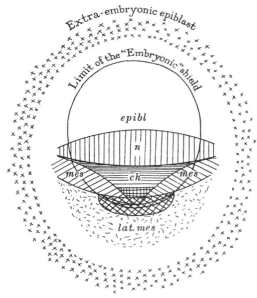

Fig. 21. Presumptive map of Reptiles. *ch, n, mes,* as usual; *lat.mes,* lateral mesoblast; *epibl,* epiblast. Cross-hatching: the prechordal plate. Oblique cross-hatching: the entoblast, with indication of the blastopore. From Pasteels, 1937.

We shall stumble on the same difficulty in the case of **Birds,** now to be studied. Leaving aside the migration of entoblastic cells, the general formative movements begin only, in chick and duck eggs, after a few hours of incubation. They are then of an extraordinary activity (fig. 22). They can only be clearly understood[1] if sufficient attention be paid to the change of the whole

[1] Pasteels, 1937, p. 402. This author has convincingly demonstrated, in my opinion, that the "Polonaise movement" described by Gräper does not really take place in development. The illusion resulted from a misunderstood combination between the general change of shape of the blastodisc, and the migration of individual territories.

shape by simultaneous extension of the two primary strata, a process clearly revealed by Bismarck brown marks applied on the vitelline membrane. The blue marks placed on the superficial

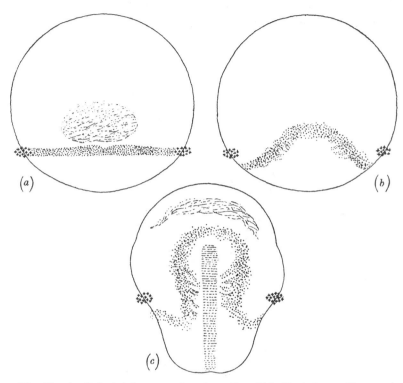

Fig. 22. A vital staining experiment on the chick blastoderm. The *area pellucida* only is represented. Small crosses: Bismarck brown marks on the vitelline membrane; firmly dotted: the Nile-blue mark; lightly indicated, the entoblastic thickening (Entodermhof), in the centre of (*a*) and the anterior part of (*c*). (*a*) The fresh mark, placed after 10 hours of incubation. (*b*) 2 hours later; note the curvation of the blue mark and its displacement relative to the brown reference marks. (*c*) $8\frac{1}{2}$ hours later ($20\frac{1}{2}$ hours incubation); curvature in Ω and appearance of the condensed primitive streak. From Pasteels, 1937.

layer (primary ectoblast) reveal then the concentration of the postero-lateral cells towards the primitive streak, their invagination, their lateral—and forward—migration between the two layers, the antero-posterior stretching of the axial material, with the backward displacement of the Hensen's node. In a word, the vital staining method, accurately checked by a study of serial sections, demonstrates the typical movements of gastrulation, but shows them concentrated along the narrow zone of the streak. In this general development, the behaviour of the presumptive materials still contained in the external layer is very remarkable. When incubation starts, they are concentrated in the embryonic shield situated in the dorsal quadrant of the blastodisc. The neural, chordal, prechordal and somitic areas, taken together, draw a kind of Napoleon's hat, put on the lateral mesoblast (fig. 23). The relative position of the notochord is still the same as in Amphibians, Fishes and Reptiles. Incubation has, after some 6 to 9 hours,[1] a double result. The primary ectoblast—soon followed by entoblastic elements—extends through its whole periphery over the enlarged *area opaca*, while the cells of the embryogenic material undergo a convergence, combined with some extension, towards the median plane (fig. 23, *b*). The neuro-chordo-somitic "hat" becomes folded in its middle. In that point, the heaping up of the lateral mesoblast makes the primitive streak apparent. Immediately thereafter, the simultaneous changes of shape in both layers begin to give the blastodisc a racket-like shape. The anterior part of the lateral mesoblast invaginates actively and the "two-horned hat" becomes laterally closed on the narrow primitive streak (fig. 23, *c*). On each side, the future somites lie, turned through a right angle with respect to their primitive orientation. Before these and the axial material of the streak the tiny chordal cap is placed, which Pasteels, after many attempts, succeeded in colouring selectively. The prechordal mesoblast is already invaginated (fig. 23, *c*, right side) and lies in front of the so-called head process. The next step brings the simultaneous invagination of chorda and somites. While they are sinking in, the latter rotate for a second time and thus regain their primitive orientation. Finally, the powerful cephalo-caudal stretching of the chordal material enters into action. It pulls the axial portion of the somites backwards and these, by a last rotation, attain their definitive transverse orientation. At the same time, invagination being at its end in

[1] Gräper, 1937; Twiesselmann, 1938.

the anterior territory, the fore-part of the primitive streak fades
away. Hensen's node seems to move backwards, and the same

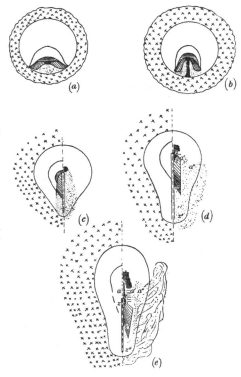

Fig. 23. Scheme of the presumptive areas of the chick blastoderm at characteristic stages of primitive streak formation (a, b, c, d), and during the
backward movement of Hensen's node (e). Stippled: the lateral mesoblast;
cross-hatched: prechordal mesoblast; white crescent: neuroblast; horizontal shading: chorda; oblique shading: somites; small crosses: extra-
embryonic ectoblast. (a) Before incubation; (b) 8 to 10 hours of incubation;
(c) 15 hours of incubation; (d) 24 hours of incubation; (e) 36 hours of incubation. In (d) and (e) the letters a, a', a'' and z, z', z'' help to show the
intensive extension of the chordal material, (a–z) as compared to the somitic
(a'–z') and lateral (a''–z'') mesoblast. From Pasteels, 1937.

process is repeated at still more caudal levels. The formation of
the caudal bud will be briefly considered later, for all Vertebrates
together.

Comparing Reptiles and Birds we have to note, besides the evident general similarity, certain differences of dynamic order. The entoblast of tortoises invaginates in a plain mass and through a real blastopore; in the chick, cells insinuate themselves, like patrolling soldiers, into the depth: the activity is the same, only more dispersed in the second case. Invagination of the chordo-mesoblast is realized, in Reptiles, uniformly and synchronously all around a visible blastopore. In Birds, it occurs gradually, the lateral mesoblast ahead, followed by the presumptively lateral part of the somites, then by their axial part, and by the chorda. But in spite of the blastopore being stretched and shut in a sort of raphe, the performance accomplished by the cells remains in fact identical. Once more, it is evident that the blastopore must not be taken in its etymological sense of primitive mouth. Neither the orifice, its form, nor its lips have any intrinsic importance. The sole significant fact is the migration actively realized by the organ-forming cell groups. At the end of cleavage, the configuration of these is indeed quite similar in both groups of Sauropsides. The mosaic of presumptive areas is absolutely identical, only their proportions are slightly modified. Movements in Reptiles will occur almost simultaneously, while in Birds they will be scattered over a long period of gastrulation.

If we extend the comparison to Anamniotes, we observe a pertinent similarity of the maps (fig. 24) in regard to the position of chordal, neural, somitic and coelomic materials. The transition is evident if the yolk-entoblastic mass of an Amphibian egg is supposed to be reduced to the small entoblastic anlage of a Reptile. The ring of the marginal material shrinks then around the blastopore, and takes on the exact disposition described in the case of the tortoise. A strong antero-posterior stretching, with accentuation of chronological differences, allows us to pass to the case of Birds. In relation with this view, the situation of the primordial gonocytes is in complete agreement. In Am-phibians, their material is primitively located at the vegetative pole, and migrates, during cleavage, towards the blastocoele floor.[1] We find them in tortoises lying in the blastoporal plate, just at the place that would be expected on the previous hypo-thesis of an entoblastic reduction. In some other Reptiles, and

[1] Bounoure, 1934, 1937.

in Birds, they are found at later stages located in a crescent which is often anterior to the embryo, but which nevertheless belongs to the entoblast. This means that they must have been concealed, before the formative displacements, in the entoblastic area of the blastodisc, and have migrated forward with the elements of the internal layer.

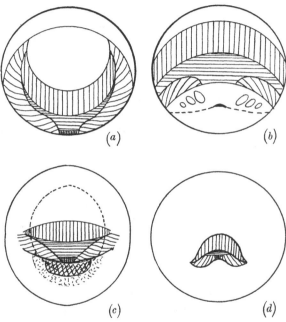

Fig. 24. Comparative maps of the main anlagen in: (a) Teleosts (trout); (b) Amphibians (toad); (c) Reptiles (tortoise); and (d) Birds (chick). Vertical shading: neural area; horizontal shading: chorda; oblique shading: somites; squares: prechordal entoblast; oblique cross-hatching (in (c)): entoblast; dotted (in (c)): lateral and extra-embryonic mesoblast.

It must however be recognized that the hypothetical reduction of the entoblastic area present in an Amphibian germ, and the actual dissociation between entoblast and the main mass of yolk, do not resolve the problem of genetic connection between the Anamniote germ and that of Sauropsides. The mesoblastic ring, indeed, would then become closed around the entoblastic anlage, but this change would not explain why in Sauropsides the epiblastic territory now surrounds the other ones as a kind of belt, nor why this epiblast gradually stretches on the yolk mass, although in meroblastic Fishes the edge of the blastodisc constantly preserves the value of a blastoporal lip. To interpret this respectively intradiscal and peridiscal situation of the blastopore in Amniotes and Anamniotes, we shall have to examine the results of experimental analysis concerning the first steps of Vertebrate development. This will reveal to us what is the actual, tangible reality enclosed in the presumptive maps and formative movements, and afford us a general representation of the organization of the cell as a germ. We shall thereby be able to fill the gap between Anamniotes and Amniotes and to understand the morphological meaning of the foetal membranes, which have been the indispensable basis for the appearance and evolution of placental Mammals.

Before entering the field of experimentation, our descriptive sketch must be completed by some remarks concerning the **caudal bud.** In all Vertebrates it has, for a long time, retained attention because of its poor structure, its numerous cell divisions, and the sudden appearance, level by level, of the various organs. Do these phenomena indicate a special mode of morphogenesis, as was suggested by Holmdahl[1] in the case of Birds, with much appearance of truth? Such a hypothesis seems to be exaggerated. Accurate studies combining, as it is always necessary, vital staining and serial sections, have been conducted on the trout, the axolotl, the tortoise and the chick.[2] They agree in revealing, in the caudal region, the same formative movements as in the main body. But the cells are small, the processes are rapid and the masses tiny. Everything is quite different from the clear-cut disposition in the classical three layers. In the modern dynamic conception of development, the distinction between those three layers is, however, deprived of any intrinsic importance. Their occurrence in the main part of the body has the value of a tactical manœuvre, most conspicuous in the initial large formative movements. But their disappearance, where the hind-part of the body

[1] 1935, 1936.　　　　[2] Bytel, 1931; Pasteels, 1936, 1937.

is concerned, does not mean any change in the essence of the process. The homogeneous aspect of the caudal bud has not, from that viewpoint, more importance than the same feature of the primitive streak. The confusion of materials long asserted to exist in the latter embryonic formation is well known, to-day, to be a pure illusion, and the same will probably be demonstrated sometime for the caudal bud.

CHAPTER V

The rôle of induction in Vertebrate development

The morphological features of development have a **signification** far surpassing the descriptive point of view. Each one of the various areas, the boundaries of which have been determined by the vital staining method, and which extend inside to a certain depth, has its particular activity. Its cells change their shape, and move in order to get a definite position. Their differential mode of manœuvre is itself evidence of individuality. The relative fixity, among the various classes of Chordates, in the distribution of the organ-forming territories at least points to some constancy in the factors responsible for morphogenesis. To discover these, to make precise the extent of their variations in the different types of germs, would be a worth while contribution to modern Embryology. Let us try to appreciate, in relation to that far-reaching purpose, the actually available experimental facts.

The **ideal solution** of the morphological problem would be a continuous series of causes and results leading from the mature but unfertilized egg to the well-formed embryo, which is already endowed with brain and spinal cord, rudiments of the sense organs, muscles and heart ready to contract, a set of pronephritic tubules, etc.

It is a fact, established by numerous grafting and isolation experiments, that these organs, when they arrive nearly at the end of their purely constructive, prefunctional period, exhibit a very definite individuality. The conditions of their final detailed complication are, however, not yet fully achieved, and many interactions and correlations are still functioning inside limited territories or between distinct regions. But the main organs undoubtedly possess their capacities of preserving their characteristic form, of elaborating cytoplasmic differentiations, and of effecting specialized physical and chemical work. It is evident that such a state of affairs has been gradually acquired and consolidated. In the case of most Vertebrates, determination can be traced step by step, in a retrograde analysis, into the apparently simple but so intimately intricate events of neurulation and gastrulation. In Prochordates, the so-called restriction of potencies happens somewhat sooner, but not so much so as a current view asserts. Anyway, *Amphioxus* and Ascidian eggs, in so far as they have been studied, represent, to a certain extent, a different type of organization, and we shall do better to postpone specifying these differences until aware of the main factors operating in Vertebrate development (p. 103).

The most prominent discovery, in that field, is indeed **embryonic induction,** i.e. the elucidation of the conditions provoking the transformation of ectoblast into neuroblast and epiblast. The admirable researches of Spemann and his co-workers have demonstrated that the dorsal accumulation of cells (p. 26), with all its remarkable consequences, is due to an influence exerted on the external layer by the roof of the archenteron, i.e. the prechordal ento-mesoblast and the chordo-mesoblast. It is not excluded that the presumptive neural area has a slight predisposition to its normal fate, but this favourable property is certainly not indispensable to neuralization. In any case, except in some precocious explantations,[1] no formation of a neural plate can take place, and no differentiation of neural cells can arise, without an external impulse normally delivered by the invaginated material of the dorso-marginal zone.

Few biological theories enjoy a stronger justification than that of the origin by induction of the neural organ, especially in the case of Amphibians. Already appearing and germinating in operations which involved dividing young and advanced newt gastrulae by sagittal and frontal ligatures, that fundamental concept was finally formulated by the celebrated embryologist Spemann as a consequence of Pröschholdt's experiment (1921) in which the blastopore lip of *Triton cristatus* was implanted on

---

[1] Holtfreter, 1931.

the ventral side of a *Triton toeniatus* gastrula and induced there a secondary embryo by a thorough reversal of the fate of the influenced materials. That stroke of genius received ample confirmation, and it acquired a still greater significance thanks to a long series of experiments: exchanging various small territories, and thereby demonstrating the mutability of ectoblastic areas;[1] grafting blastoporal lips between genera with positive induction results;[2] inserting various parts of the marginal zone in the blastocoele, inductions being obtained only with the dorsal ones;[3] sticking together dorsal halves of newt gastrulae and thereby producing *duplicitates cruciatae*;[4] detecting the inductive capacity of the archenteric roof, by grafts[5] and by extirpations;[6] observing the effects of an exactly limited median deficiency;[7] proving the inductive efficiency of a blastoporal lip grafted in a previously isolated ventral half of a gastrula;[8] testing most methodically the properties of the neural material at close stages;[9] inverting square pieces of ectoblast cut on the border of the presumptive neural territory, to change epiblast into neuroblast and vice versa;[10] comparing the inductive power of the chordal and somitic part of the archenteric vault, the balance being favourable to the former;[11] extending to Anurans the inductive processes discovered in Urodeles;[12] performing aseptic explantations in a culture medium, which did not yield active neural organization of purely ectoblastic material, and only occasionally some neural differentiation;[13] provoking, by exo-gastrulation, an integral separation of ectoblast and other materials, the former simply showing the epibolic surface augmentation, combined with some dorsal convergence;[14] forming xenoplastic combinations between a chordo-mesoblastic fragment of one species and an ectoblast fragment of the other, and thus obtaining at the same time induction in the ectoblast and curious alterations of the graft, which will be considered later;[15] grafting in successive stages

[1] O. Mangold, 1923.   [2] Geinitz, 1925.
[3] Bautzmann, 1926; see also Vitemberger, 1932; Schotté, 1936.
[4] Wessel, 1926.   [5] Marx, 1925.
[6] Lehmann, 1926.   [7] Spemann and Wessel, 1927.
[8] Bautzmann, 1927.   [9] Mangold, 1929.
[10] Lehmann, 1929.   [11] Bautzmann, 1929.
[12] Schotté, 1929.   [13] Holtfreter, 1931, 1934.
[14] Holtfreter, 1933.   [15] Holtfreter, 1935, 1936.

the right and left halves of the dorsal lip, and thereby testing
the progress of its determination in the lateral direction,[1] etc.[2]

To take at least one concrete example, and a not too hackneyed
one, let us consider the **experiment of translocation,** the knowledge of which will lead us to an exact comprehension of the state of
organization at the very beginning of morphogenesis. The egg of

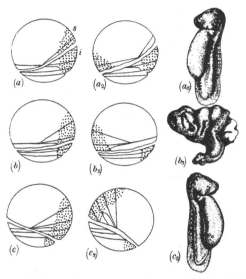

Fig. 25. Scheme of the translocation experiment. (a), (b), (c) Left side view
of the *Discoglossus* egg with the general map of anlagen and the three types
of section dividing the vital mark in two parts. (a_2), (b_2), (c_2) The same germ
after the 180° rotation of the animal part. (a_3), (b_3), (c_3) The embryos
obtained, with the parts of the marks which are still visible. Heavy dots:
the still superficial stained material; empty circles: the invaginated stained
material, visible through the skin.

Discoglossus pictus, a Mediterranean Anuran, to which Wintrebert
has called attention, can be most readily liberated from its
envelopes, and resists marvellously, in certain definite conditions,
operative procedures which would seem unreasonable if performed
on other Amphibian germs. From the plain blastula stage to

[1] Mayer, 1935.
[2] For a detailed account of these researches, till 1935, see Dalcq, 1935:
L'Organisation de l'œuf chez les Chordés.

the rounding of the blastopore, any cut may be performed without any serious loss of cells. If a platinum wire is used, the only injured elements are the most superficial ones, cemented together in the black pigment coat. The deeper cells are perfectly round, very slightly adherent, and they are gently separated by the thread without being injured. Except in cases where some material leaks from the wound, owing to some irregularity or folding of the coalescing edges, extensive transpositions of large areas may be performed without any appreciable deficit. The interest of these preliminary remarks will appear later.

Another advantage of these eggs is a very distinct grey crescent characterizing the dorsal subequatorial territory. The blastoporal groove sinks into the lower part of this crescent, just above some rather big, yolk-charged blastomeres. Let us now place (fig. 25) a large Nile-blue mark on this dorso-marginal territory of a jelly-less blastula or young gastrula. Removing then the last chorionic membrane, we cut the egg latitudinally, somewhere in the equatorial region, and separate the animal and vegetative halves. The former is turned through 180° around the main egg axis, and again placed, for healing, upon the latter. As a result, the upper part of the mark will be transferred on to the ventral side. Its lower limit will indicate the ventral level of coalescence, and the upper limit of the dorsal part will give the same indication on the other side. Consequently, the level of the cut may be retrospectively ascertained at any moment, even on sections, granted a stain-preserving method[1] has been used.[2] If we do not take into account, for the moment, certain interesting oblique or irregular cuts, the results, concerning induction, clearly fall into three groups. Operations performed at a high level, intersecting only the presumptive neuroblast and epiblast, would remain unnoticed in the structure of the embryo, were it not for the half marks (fig. 25, a). If the cut is somewhat lower, on the edge of the black cap or in the grey pigmented area, the chordomesoblast is divided into two parts, the upper one being brought ventrally; double embryos are obtained, which present in most

[1] Lehmann, 1929.
[2] Töndury, 1936, has made on the urodele gastrula experiments of partial translocation of the inductive centre to more animal regions, with interesting results for the dynamics of gastrulation.

cases a spina bifida on the originally dorsal side, and other
particularities of structure to be considered later (fig. 25, b). In
cases where the section attains the level of the blastoporal groove,
or lower, a simple embryo arises according to the situation of the
ento-chordo-mesoblast, now transferred to the originally ventral
side[1] (fig. 25, c).

In spite of considerable difficulties, the inductive origin of the
neural organ has been ascertained in some **other classes of
Vertebrates.** In Fishes, *Salmo irideus*,[2] *Perca fluviatilis* and
Fundulus heteroclitus[3] have been submitted to conclusive experiments. The most typical one consists in removing, in the
trout egg, the just invaginated material through the roof of the
subgerminal cavity and transferring it under the ventral edge
of the blastodisc (fig. 26).[4] In Birds, Waddington[5] succeeded in
separating the two layers of the shortly incubated blastodisc,
and in cultivating two external layers of primary ectoblast, stuck
together with opposite orientation. He saw that the posterior
region induces a neural organ in the epiblast with which it comes
in contact. A very interesting mode of induction is also observed
if the two layers of the didermic stage are separated after a few
hours of incubation and the entoblast rotated 180° relatively to
the primary ectoblast: the dorsal part of the internal layer
induces an embryonic system in the normally indifferent part
of the external layer. In Fishes, Amphibians and Birds the
translocation experiment is thus able to yield double embryos.
Interesting attempts have also been performed to extend the
demonstration of inductive processes to Mammals;[6] if still not
absolutely conclusive, their results are certainly not opposed to
such a generalization.

These fundamental relations being settled, our attention
naturally turns to the elements which are **responsible for this**

---

[1] Dalcq, 1933, 1934, 1935.   [2] Luther, 1936.
[3] Oppenheimer, 1936.
[4] In other experiments, Luther effects a 180° rotation of the primary
ectoblast relatively to the invaginated material, and describes an induction
by the dorsal part of this deep layer.
[5] 1932, 1933.   [6] Waddington, 1937.

neural induction. They form a real complex of invaginating dorso-marginal materials and simultaneously perform two distinct tasks: from their own cells, they build up some typical organs, while they influence by induction the overlying ectoblast. When the primary inductive complex, represented by the young blastoporal lip, is grafted on the ventral side of an host, a third effect takes place. To cause an embryo-like structure to appear, the host mesoblast and entoblast must undergo a deep remoulding. The former—the fate of which was only to build up the coelomic linings, mesenchyme and blood—joins the grafted chordo-meso-

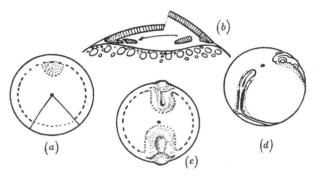

Fig. 26. Translocation of primary inductive material in a young trout gastrula. (*a*) Schematic view of the blastodisc with the limit of the sub-germinal cavity (broken line), and the invaginated material (dotted), and the direction of the angular section. (*b*) The operation represented on a sagittal section. (*c*), (*d*) Two stages of the twin embryos obtained. Redrawn from Luther, 1936.

blast and co-operates to form somites and sometimes chorda; the latter, instead of being included in the intestine, completes the walls of the archenteron and is eventually changed into a pharynx. This fact clearly indicates that the conditions governing the development of entoblast and the segmentation of mesoblast can be artificially brought into being through the presence of a grafted centre; the representation which we are seeking for these processes should account for this rearrangement (p. 98). It has certainly an intimate relation with the results obtained with a graft, in the dorso-marginal region, of a small piece of ectoblast; this is readily transformed into either chorda, or somites, or pronephritic tubuli.[1] In the case of a grafted dorsal lip, the appearance of the characteristic structures of the embryonic body is thus due to a general remoulding effect exerted by a specialized, leading material. The whole process can be conveniently named *Evocation*, as suggested by Waddington. This appeared to Spemann as a fact of typical organization. The graft capable of such action was by him christened an *organizer*, and the same material, in its position, an *organization centre*.

Many attempts have been made to define the nature of its influence, and have led to new discoveries. The inductive power may be transmitted to an ordinary piece of ectoblast inserted for some hours in the blastoporal lip,[2] or in the chordo-mesoblast of a blastula, before any invagination.[3] The same property also exists, concealed, in the neural organ: young ectoblast, submitted to its influence, undergoes the typical transformation.[4] Although the capacity of induction is clearly limited in the normal young gastrula or in the blastula to a definite territory, a similar inductive power can be found not only in blastoporal lips or primitive streaks[5] of other species, not only in the neural organ[6] which is itself induced, but also in most varied materials, generally of animal origin. In nearly all cases where living material seems to be inactive, it becomes efficient by boiling, freezing, alcoholizing, etc.[7] A special mention must be made respecting extracts of blastoporal lips.[8] Again, a certain number of pure organic chemicals have been found to provoke more or less typical inductions. What substance really acts in normal development remains however uncertain. In spite of the tenacious attempts of the Freiburg[9] and Cambridge[10] laboratories, the

[1] O. Mangold, 1923.   [2] Spemann and Geinitz, 1927.
[3] Raven, 1935. This shows that the inductive power is certainly not a consequence of the invagination in itself, as Wintrebert assumes it to be.
[4] O. Mangold and Spemann, 1927.   [5] Hatt, 1934.
[6] Morita, 1936.   [7] Holtfreter, 1933.
[8] Waddington, Needham and Needham, 1933; the same, with Nowinski and Lemberg, 1934, 1935.
[9] Fisher and Wehmeyer, 1933, 1935.
[10] A recent attempt has been made to test the respective importance of desmo- and lyo-glycogen in causing evocation and individuation. Grafting experiments have indicated that desmoglycogen only provokes massive neural inductions, i.e., apparently evocation without further individuation (Heatley, Waddington and J. Needham, 1937). However, in the Axolotl blastula, glycogen is more abundant at the animal pole than in the vegeta-

chemical problem of induction still requires further investigation. The orientation of these researches could be helped by a better knowledge of the general nature of the morphogenetic factors; this is a supplementary reason for pursuing the present inquiry, which, I hope, will not be found too deceptive. An important hint is also provided by the enlargement and eventual duplication of the organization centre, when submitted to a lateral gradient of temperature.[1] Respiratory metabolism may be suspected to play its part in the process. Ectoblastic ventral material, normally deprived of any evocation power, acquires the ability of inducing neural structures if previously submitted to a respiratory catalyst, methylene blue.[2] On the other hand a larger CO_2 elimination and a higher respiratory quotient have been directly demonstrated to characterize the blastoporal region of the Amphibian egg.[3] Very recently, measurements could be made by an analysis of the gas exchange taking place in a dorsal, ventral or lateral hemisphere of a perfectly intact gastrula.[4] They indicate, for the dorsal side, an excess of O_2 absorption amounting to about 45 per cent. Such a result could of course be foreseen on account of the prevailing organizing activity of the dorso-marginal region. But if future improvements allow measurements over shorter periods, especially at young stages, it could finally be proved that respiratory metabolism plays an effective rôle in the production of the morphogenetic conditions, which is a most tenable assumption.

Putting together the actual data relative to induction, we may assert that this process is essentially chemical. Purely physical particularities can be dismissed: the actual direction in which formative movements are effected[5] is not a cause of neural differentiation. Neither can the effects of the sap contained in the blastocoele be considered as the epigenetic origin of the inductive property.[6] The reaction happens as if some efficient

tive region. During gastrulation, a decrease of about 30 per cent occurs in the total glycogen of the invaginating material. "It is unlikely that desmoglycogen is identical with that fraction of glycogen to which the evocator is attached in the cells" (Heatley and Lindahl, 1937).

[1] Gilchrist, 1928; Castelnuovo, 1932.
[2] Waddington, Needham and J. Brachet, 1936.
[3] J. Brachet, 1936.   [4] J. Brachet and Shapiro, 1937.
[5] Goerttler, 1927.   [6] Wintrebert, 1935.

compound or substance were present in the blastoporal lip, the young primitive streak, the neural plate and tube, and in exceedingly varied adult organs, either when living or when killed by certain procedures. For the sake of brevity, I feel it convenient to refer to the chemical conditions responsible for induction in the normal egg—and probably in most experimental conditions— as "organizin", a designation not prejudging in any way the origin, the nature, or the mode of action of this or these products.

It must finally be remarked that induction strongly affects the **auxetic surface process** (p. 28). The primary induction centre clearly favours the growth of internal surfaces necessary to invaginative deformation. When the roof of the archenteron attains a sufficient contact with the ectoblast, a similar activity occurs on the underside of the ectoblastic layer. Later on, an analogous, but more localized, process occurs between the neural crests and epiblast, out of which placodes are induced.[1] Consequently, the organizin can be imagined as one or several substances transmissible from a cell layer to another one by intimate contact, through the cortical film, and producing, among other possible effects, an auxetic surface activity. The resulting change of shape could thus be attributed to a lowering of the surface tension[2] or some more active, true growth of the surface film.

[1] The demonstration is complete for the large placodes of the nose, eye and ear, but not for the small placodes of the cranial ganglions.

[2] This view is also expressed by T. H. Morgan (1937) concerning the invagination in the gastrula. An analogous suggestion has been presented, on a mathematical basis, by Rashevsky (1933, p. 186), who considers the organizin as "excitatory substances".

CHAPTER VI

Regional organization in the Vertebrate germ, after cleavage

The general process of "Evocation" is always in normal development, and often, in experimental conditions, associated with the arising of local inter-related differences, especially in the external and the middle layers. This differentiation *sensu largo*, "Individuation" in Waddington's terminology, is apparent from the very first indication of the normal neural plate. The difference of width between its anterior and posterior parts clearly foretells the important **distinction between head and trunk,** and all the correlated complications.

The conditions governing this new process can be analysed in certain graft experiments. The young blastoporal lip appears, in normal development, as the probable inducer of cephalic structures, while the later invaginating material seems responsible for the trunk organization. Head and trunk organizers actually differ in their inducing power, but the region of the host where the reaction occurs is not without importance. The induced heads are not so typical, when obtained in a trunk region of the host, and a trunk organizer is able to induce a head if inserted in the cephalic area.[1] Grafting an older blastoporal lip in the place of a young one causes a remarkable slenderness of the brain, which is not capable of forming optic vesicles; but the inverse operation does not raise the trunk structure to a cephalic type[2] (fig. 27).

What is the intrinsic value of the methodological distinction introduced by these experiments? How much is the head organizer **qualitatively different** from the trunk-inducing material? Is not the same agent able to produce both types of structure according to its concentration in a definite locality of the germ?

[1] Spemann, 1931.　　　　　[2] Hall, 1932, 1937.

The presumptive significance of the invaginated material furnishes a first and important indication. The head organizer contains prechordal ento-mesoblast and a large amount of presumptive chorda. But the vanishing of any cephalic effect, when these materials are inserted in the dorsal lip of a circular blastopore, at least indicates that they are not capable, by their own exclusive forces, of developing into cephalic structures. On the other hand, if a qualitative difference was implied, the inferior and central parts of the organizing centre would, in the translocations, show a greater inductive power than the higher and more lateral regions of the chordo-mesoblast. In fact, the induction observed

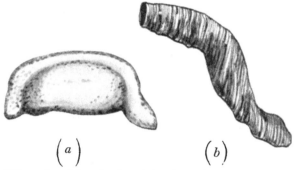

$$\left(a \right) \qquad \left(b \right)$$

Fig. 27. Effect of grafting the dorsal lip of an advanced Triton gastrula in the place of the dorsal lip of a young germ. (a) Aspect of the embryo, from the right side. (b) Plastic reconstruction of the brain. Redrawn from E. K. Hall, 1937.

on the primitively dorsal side, in a case of low translocation, may be equivalent to the one obtained on the primitively ventral side in a case of high translocation (fig. 28): in both double embryos, the smaller part of the inducing centre induces a spinal cord with, anteriorly, some nervous ganglia and one or two auditory vesicles. This result has led me to suppose that quantitative conditions may be very important for the type of induction, considered in its general features. As many authors have indeed suspected, the mode of gastrular invagination has a real importance for forthcoming induction, which is therefore intimately related to the formative movements. Translocations clearly show that lower and central parts of the dorso-marginal crescent differ consider-

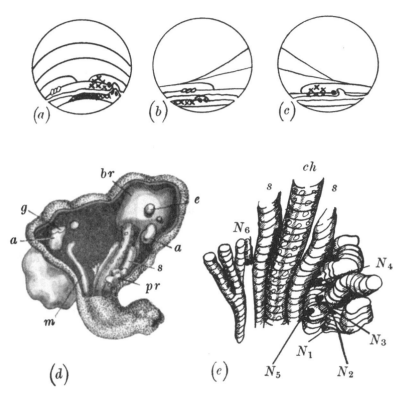

Fig. 28. Translocation experiments in the young *Discoglossus* gastrula. The cut was made just in the blastoporal lip and it extended higher on the right side than on the left. (*a*) Dorsal view indicating the level of the section. The limits of the presumptive areas (cf. fig. 12) are drawn in plain lines, with the three right (black) and left (white) nephrons. Crosses indicate the blue guide mark. (*b*) The same egg, from its right side, after the rotation of the animal cap. The lower part of the mark is visible; the three left nephrons have been brought in the vicinity of two right ones. (*c*) The same operated egg, from its left side, showing the upper part of the mark and one right (black) nephron. (*d*) Schematic representation of the double embryo, its primitively dorsal side being turned left. The dorsal system, induced by a part of the presumptive head organizer, is only formed of truncal organs and an auditory vesicle. The other system is complete. *a*, auditory vesicle; *br*, brain; *e*, epiphysis; *g*, ganglion; *m*, spinal cord; *pr*, pronephros. (*e*) Reconstruction of the pronephros of the double embryo, from the coelomic side; N_1 to N_6, the nephrostomes, disposed according to (*b*) and (*c*). *s*, somites; *ch*, chorda.

ably in their dynamic capacity from the higher and lateral ones. So long as the cut is not made above, let us say, the half of the organization centre, the dorsal part of the blastopore is not able to close; the archenteron remains short and dorsally open (fig. 28, *d*), while the ventral material succeeds in forming an embryonic complex possibly deficient in its head, but fairly normal in its caudal portion. That means that the upper and lateral elements of the dorso-marginal zone possess, in a higher degree, the potentialities of convergence and extension, and that in the natural course of events they are responsible for the main elongation of the embryo, and for the formation of its dorsal region. From these regional conditions and from the nature of the graft itself a conflict may arise, causing a head organizer to elongate more than normally, and diluting, it might be said, its inductive power, or, on the contrary, reducing the tendency to extension of a trunk organizer, with the opposite consequence. But the influence of the region where the graft is made must not be lost sight of. The head territory apparently has a tendency to raise the inductor to the cephalic level; and the trunk territory seems not to affect the individuality of the head organizer. A satisfactory representation of the regional organization has to account for these facts.

If we wish to pretend that we are able to explain, with our logical faculty as a tool, the process of Ontogenesis, we must not only consider the inductive capacity of the dorso-marginal material, but also the gradual complication of its own structure. The organ-building differentiation implies, as an inexorable logical necessity, some kind of **prestructural differences.** These cannot be conceived to exist in the nuclei, which have been demonstrated to be equipotential, at least in the young stages;[1] and the results of the innumerable grafting experiments could scarcely be compatible with a nuclear mosaic. Regional particularities, with a cytoplasmic basis, must consequently be searched for. They are known to exist a short time before the individualization of the organs, when determination has been settled. Before that period, there can be no question of conspicuous differences in the ectoblast

[1] Spemann, 1928.

region, the evolution of which is subordinate to induction, reserve being made again of some slight predisposition of the neural area. For the rest of the germ, it is important to decide whether diversification appears only when the materials have attained their characteristic location, or before that time. The significance of the formative movements is intimately connected with this question.

The term mosaic, with its rigid background, is however to be avoided, for the following reasons. When large pieces of the organization centre are removed, the embryo is sometimes deprived of such an important organ as the notochord,[1] but in other cases an apparently identical operation does not cause any qualitative deficiency. Gastrulae may be cut along the median sagittal plane, or in its immediate neighbourhood, and their halves are able to form dwarf embryos with almost perfectly symmetrical organs.[2] Lateral halves of the young blastoporal lip, when grafted in favourable conditions, build a median chorda and bilateral somites.[3] When a young gastrula is submitted to the influence of LiCl, a variable amount of its presumptive chordal material is transformed into somitic mesoblast.[4] According to Holtfreter,[5] a piece removed from the presumptive chordal field of the newt yields chorda, muscle cells and pronephros together. The same is true for a more lateral fragment, of a mainly somitic significance. Chordo-mesoblast consequently appears to be formed by interchangeable elements. But other experiments reveal something more. By xenoplastic experiments between newt and frog or toad, the same author has recently demonstrated that the material of the blastoporal lip is able to develop partly into ectoblast and all the ectoblastic derivatives.[6] Other studies concerning the cultivated blastoporal lip of the newt[7] have given rise to the new idea of a structural complexity nearly proportional to the number of fragments cultivated together. The area to be studied is removed from just above the blastoporal groove and contains presumptive prechordal ento- and mesoblast and the lower part of the chordal anlage (fig. 29). By a slight variation

[1] Dalcq, 1933.   [2] Cf. Schmidt, 1930.
[3] Mayer, 1936.   [4] Lehmann, 1937.
[5] 1933.   [6] Holtfreter, 1937.
[7] Lopaschov, 1935.

of certain of the preceding results, one or two such pieces only yield muscular cells, but three or four together already build up notochord and muscle; if a mass is composed of six to ten coalescing fragments, it shows a true segregation of an ectoblastic part, with the consequent inductions, and morphogenesis proceeds to a remarkable extent. These alterations of chordo-mesoblast into ectoblast are also to be observed in a few of my own translocation experiments. As a result of a section—followed by the

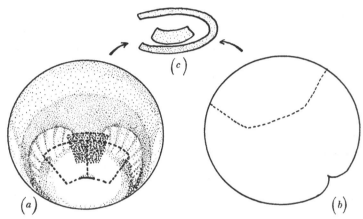

Fig. 29. Scheme of the xenoplastic Chimera in Holtfreter's experiment. (a) Presumptive map of the newt, with indication of the explanted areas (interrupted line). (b) Gastrula of an Anuran, with the ectoblastic territory to be isolated. (c) The piece of the inductive centre is placed inside the ectoblast (sandwich compound). In (a) the dotted area indicates the territory explanted in Lopaschov's experiment. Redrawn from Holtfreter, 1935; (a) modified.

180° rotation—certainly made in the chordo-mesoblastic field, the dorsal material shows some reluctance to invaginate, owing to the absence of the normal push from the trunk organizer. It causes a part of the chordo-mesoblast to remain on the surface, and to undergo induction to neural tissue. It is obvious, from the three last results, that the material of the primary inductive centre manifests, when physiologically more or less isolated, two main tendencies. The first causes the building of chorda, somites and eventually nephrotomes, and falls into line with the normal development of the mid layer. The second is of a more dynamic, organizing nature: some favoured cells succeed in creeping under the others and submit these to the control of induction.

Each one of the above-mentioned facts poses a new equation for the organization problem. They are mostly remarkable by a sort of amplification of the normal activities. They do not, however, contradict in any way the idea of some pre-existing structure governing, in the normal germ, the intensity and orientation of the formative movements, the nature of the forthcoming differentiations and the local modalities of induction. They only indicate that these properties are very delicate, almost labile, and consequently can only be discovered by methods which disturb less roughly the organized system of the germ.

Two operative procedures fulfil this condition. Local deficiencies of the marginal zone, caused by delicate cauterizations, have been shown, in *Pleurodeles* blastulae,[1] to leave definitive traces in the organs of the hatched embryo, the size of the corresponding somites or of its pronephros being notably reduced. This is a strong indication in favour of the existence of regional fields prior to gastrulation. In spite of the above-stated fact that most excisions of the blastoporal region are nicely compensated, an appropriate removal of a horizontal cell band out of a *Discoglossus* blastula causes the development of a microcephalic embryo in which the chorda is entirely lacking.[2] Still more detailed evidence relative to regional properties is provided by latitudinal and oblique translocations of the blastula and gastrula, again in *Discoglossus*. When the double embryos obtained by this operation (p. 52) are thoroughly studied with an appropriate graphic method of reconstruction,[3] the distribution of the organs met with acquires a double meaning. To a great extent it provides additional evidence in favour of the well-known recompleting of organizing parts, and in favour of the dependent inductions, with some additions to be considered later. But it also reveals a clear tendency of the marginal material to follow its normal lines of development, in spite of an abnormal location in the whole. For instance, a cut having been made high dorsally, in the neural

[1] Suzuki, 1929.
[2] Dalcq, 1933. Chorda excisions were also performed at a later stage by Lehmann, 1926.
[3] Lison, 1936.

area, but low enough ventrally, to divide the ventral mesoblast, a wedge of mesenchyme is found penetrating in the posterior brain vesicle (fig. 30). In other cases, separate pieces of notochord are discovered, owing to the isolation of the horns of the chordal area by the experimental section. Although mutations between somitic mesoblast and chorda are frequent, the main pieces of

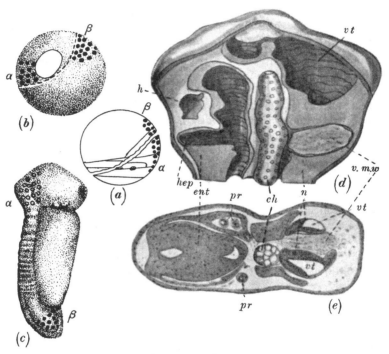

Fig. 30. Translocation experiment on a *Discoglossus* blastula. (*a*) Lateral view, from the right side, with indication of the section; the dorsal blue mark is divided into an upper and lower part, β and α. (*b*) The healed gastrula at the circular blastopore stage; the part β of the mark is opposite to α, as a consequence of the 180° rotation of the animal part. (*c*) The developed embryo, with mark α partially visible, by transparency, in the brain; mark β on the ventral posterior region. (*d*) Graphic reconstruction of the embryo supposed cut on the sagittal plane; a view on the right half, showing the mesenchyme wedge enclosed in the posterior brain region. (*e*) The horizontal section at the level of this wedge. *ch*, notochord; *ent*, enteron; *h*, heart; *hep*, hepatic recess; *n*, neural organ; *pr*, pronephros; *vt*, brain ventricle; *v.m.w*, wedge of ventral mesenchyme.

each material are located, in the double embryos, according to the prospective fate.[1] Small isolated coelomic cavities are observed and may be explained from the same viewpoint. The position and structure of the heart, median in one case, lateral in a second embryo, subdivided in a third one, suggests the idea of some heart-tendency inherent in a small anterior area of the hypomere strip. Similar observations concern the branchial anlage; asymmetrical incisions yield results that point to the existence of a specialized mesoblastic field which, in my opinion, secondarily impresses the entoblast and causes the moulding of its branchial pouches.

This group of characteristic relations already excludes the conception that the marginal zone of the late blastula or young gastrula is a homogeneous material. To the various tendencies expressed by the formative movements some very delicate properties, which can be termed *inclinations*, correspond in the intimate structure of the presumptive areas. But exactly as a young man, who was predisposed to painting, finally became Pasteur, and as many artistic or scientific vocations are daily impeded by inimical conditions, so are the inclinations of the marginal territories easily distorted into different utilizations. What a minuteness these local specifications can attain, however, is clearly shown by the organization of the primitive kidneys in the embryos resulting from the same translocated germs.

The development of the nephrotomes happens immediately after the individuation of the chorda and somites, and nobody seems to have ever looked for a precocious localization of their material. Descriptions of the numerous twin or induced embryos, obtained by various methods, generally lack any indication concerning the kidneys, and no direct experiment has ever been performed on their anlage at stages preceding the circular blastopore. In that middle phase of gastrulation, rotations of their presumptive area[2] reveal that their cephalo-caudal axis is already established, just as is known to be the case for the neighbouring material of the anterior limb.[3] What could be expected of stages preceding

---

[1] Translocations of mostly chordal material in young gastrulae of *Limnodynastes tasmaniensis* (Bautzmann, 1932, 1933) also reveal the individuality of the chordal anlage.
[2] Ti Chow Tung, 1935.   [3] Detwiler, 1933.

the crisis of determination? Would not both axial systems of the double embryos be provided with two normal sets of the three[1] pronephritic tubuli, with their open and ciliated nephrostomes? It seemed, indeed, very likely. Nevertheless, oblique cuts, similar to that of fig. 28, showed the possibility of displacing, together with the branchial mesoblastic field, the whole right or left pronephritic anlage. Integral reconstruction of numerous translocated embryos then revealed extremely curious structures. In place of the normal or slightly deficient organs which were to be expected in both embryos on the hypothesis of a regulative epigenesis, the most varied dispositions of the nephrons[2] were observed.[3] Let us describe briefly some characteristic instances. (1) One individual of the double embryo possesses the two excretory organs, with the normal number of tubes and nephrostomes, perhaps slightly reduced in size; the other individual, more or less rudimentary, has no excretory organ. (2) Each individual has only one pronephros, and the organs belong to complementary sides, one being right, the other left. Absence of one or two nephrons may occur. (3) A single, but complex organ is encountered between the two individuals, being e.g. left for one and right for the other; it shows more than three nephrostomes, five in the case represented by fig. 31. Such an arrangement results from an aggravation of the previous type, the respectively left and right organs having fused, owing to a rotation happening to surpass 180°. (4) The pronephritic material is divided into two, three or four pieces, which, taken together, sum up five or six nephrostomes. A typical case is demonstrated by fig. 28; the main embryo possesses, on its right side, a pronephros with three tubes, two blind ones and one, longer, opened by a broad nephrostome; on its left side, a complex organ, common to both embryos, has five nephrostomal openings. The cut has thus intersected the presumptive right kidney area and amputated the juxta-coelomic part of two nephrons, which has been brought on the ventral side and has become fused with the whole left

---

[1] Typical number for Anurans. In Urodeles, each pronephros has only two nephrons.

[2] The nephron is the unit formed by each tube with its nephrostome opening in the pronephritic space of the coelomic cavity.

[3] Dalcq, 1937 *a, b.*

anlage. (5) Three or four organs exist, in the best cases symmetrically placed; they number more than six nephrons; eight, nine, eleven, and in *one* simple case, twelve (fig. 32).

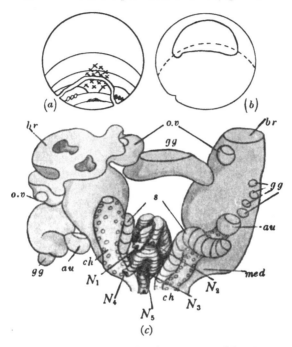

Fig. 31. Asymmetrical translocation in a young *Discoglossus* gastrula. (*a*) Dorsal view with the indication of the blastoporal lip and the presumptive areas; the three white and three black spots indicate the left and right nephrons respectively; the crosses show the reference mark. (*b*) Level of the cut as seen in a sagittal section, blastopore to the left. (*c*) Schematic reconstruction of the double embryo, especially to show the single pronephritic organ, with five nephrostomes, N_1 to N_5. *au*, auditory vesicles; *br*, brain; *ch*, notochord; *gg*, ganglion; *med*, spinal cord; N_1 to N_5, nephrons with their nephrostomes; *o.v*, optic vesicle; *s*, somites.

The arrangement which would most frequently happen, if regulative epigenesis had occurred, i.e. a double embryo with four complete kidneys, is thus absolutely exceptional and, as we shall explain, its occurrence is only possible after a well-localized cut. The series of observed structures can only be understood on the assumption of a precocious inclination of the nephrogenous material. Its anlage is known, by descriptive studies, to prolong ventrally the second, third and fourth somites. This indication allows us to draw its presumptive field on the *Discoglossus* map, in spite of the fact that a vital stain rarely reaches this somewhat

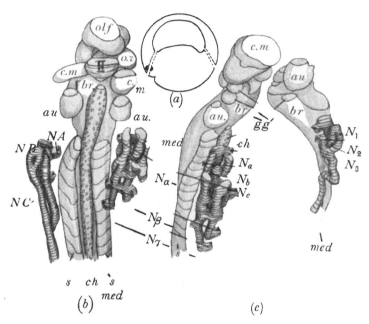

Fig. 32. Symmetrical ascending translocation in a *Discoglossus* young gastrula. (*a*) Level and direction of the cut as seen on a para-sagittal plane; blastoporal lip, left; hatched area: the pronephritic anlage. (*b*) and (*c*) Reconstruction of the principal organs and specially of the pronephroi of the two combined embryos; both reconstructions viewed from the coelomic side; *NA*, *NB*, *NC*, the three nephrons of the right pronephros of one embryo; N_α, N_β, N_γ, the same of the left; this organ is fused with the right one of the other embryo, where nephrostomes N_a, N_b, N_c are encountered on the right side, N_1, N_2, N_3 on the left. *au*, auditory vesicles; *br*, brain; *ch*, notochord; *c.m*, cephalic mesoblast; *gg*, nervous ganglion; II, infundibulum of the prosencephalon; *med*, spinal cord; *s*, somites; *o.v*, optic vesicle; *olf*, olfactory vesicle.

deeply located material. Each of the observed cases may consequently be attributed to a certain level of section relative to the presumptive pronephros. The reader can easily deduce what level it must have been for the three first items, some slight loss of material not being always avoided. Protocol notes, and also

the retrospective indications gathered from the slides thanks to
the preserved marks, agree perfectly with these inferences. But
the fourth series means indeed something more than a predisposition of the whole anlage: nephritic units, especially the parts
forming nephrostomes, have been independently left *in situ* or
translocated. Does this disjunction mean some preparation of
the forthcoming segmentation, inside the whole anlage? If such
is the case, the presumptive nephrons would constitute small
elongated pieces with a very oblique, nearly horizontal long axis,
in the prolongation of their respective somites. A latitudinal
section might indeed happen to disjoin them, sometimes with
partial injury or material loss. The possibility of splitting the
units must then be faced, and the example of fig. 32 illustrates
its reality. It also gives the clue for understanding the embryos
with supplementary nephrostomes. In fact, some of these present
differences of size are to be attributed to the chance cut. But,
owing to the orientation of the tiny presumptive nephrons, it can
be foreseen that the best opportunity of dividing each of these
would be realized with an ascending, nearly tangential cut, from
the blastoporal lip towards the dorsal edge of the blastocoele
(fig. 32, *a*). It was actually such an operation which provided
the double embryos with eleven and twelve nephrons.

The striking nature of these results makes it necessary to avoid
any exaggeration in their **interpretation**. We must be careful to
exclude any too marked, unfounded preformation. There is no
question here of an independent differentiation of the pronephritic
material. The described predisposition essentially concerns the
nephrostomes, which are the primary communications between
the coelomic cavity and the tubuli. It does not mean—these
questions are not yet settled—that the size of the organs is
limited, nor that they are not sometimes completed at the expense
of neighbouring material. Direct experiments of excision, rotation,
isolation and displacement have been performed on the field containing the pronephritic area. Some excisions have caused clear-
cut deficiencies corroborating the previous results. Grafts in
atypical situation have shown the great lability of the innate
nephrogenous tendency, of which only a tubular orientation of

the grafted cells persists. But many embryos presumed to have been deprived of one or both kidneys nevertheless have recovered a nearly perfect structure. There is no doubt that a deficiency affecting a limited lateral territory of the inductive centre is often completely compensated, thanks to a local reaction. This happens to be avoided in translocations, and this difference between large and localized operations must be accounted for in the general conception that we aim at.

The analysis of induction has caused us to consider the intimate, functional and morphological structure of the marginal material, especially in its dorsal, inducing territories. Having recognized the existence of their extraordinarily minute pattern of predisposed fields, we are led to examine again what may be their bearing on the **structure of induced organs.**

The pronephros, which has been rather extensively studied here, has no evident inductive function. My attempts to find some relationship with the head and trunk organizers, with the ear and the anterior limb bud, have so far failed. There are, however, not far from the pronephritic area, some elements quite active in induction. Their activity concerns the placodic derivatives of the epiblast, and compels us to consider again, for the last time, these very instructive translocated embryos. In a number of these, the auditory vesicles show a kind of balanced disposition: a right one in one embryonic system, a left one in the other; a right big one and a left small one in one member of the pair, and the reciprocal in the other, etc. Some embryos possess giant otocysts, or scattered minute ones. Generally speaking, results obtained with the blastulae tend to be more disordered than with gastrulae, where the cases of exact balancing are more frequent. This indicates a certain process of maturing which gradually stabilizes a special, inducing focus, situated anteriorly to the pronephros, perhaps even in the prechordal mesoblast. Some property located there—a stronger concentration of organizin?—plays an essential if not exclusive rôle[1] in the induction of the auditory placodes.

[1] An influence is probably exerted also by the rhombencephalic level of the neural organ. The localization and size of the normal otocyst probably depends on a complex situation, by a system of "crossed fires" (Holtfreter, 1935). See also Lopaschov, 1937 and Schmidt, 1937.

Anomalies are also produced, in these atypical heads, concerning the structure of the pituitary, the olfactory placodes and the eyes. These alterations appear to be the result of a variable action of the prechordal material. From this small group of important cells depends, according to other experimental evidence, the existence of one or two eyes,[1] of the two olfactory lobes, and of the olfactory placodes, subordinated to the lobes.[2] Translocations authorize us to add to these typical head organs the epiblastic part of the pituitary, induced by the prechordal entoblast.[3] In the well-considered conception of Adelmann, the non-occurrence of cyclopia is thought to be related to the symmetrical structure of the substrate, with its prechordal plate laterally flanked by the strips of prechordal mesoblast which soon form the mesoblastic cores of the mandibular arches. Further researches will probably unravel, in this intricate region, correlations similar to those known in the chordal part of the body, where chorda, mesoblast and neural tube exert reciprocal influences upon which the internal conformation of the spinal cord[4] and the localization of spinal ganglia[5] depend. The situation would be considerably clearer if an explanation were offered for the segmentation of the dorsal mesoblast in somites: still one desideratum more for the wanted theory (p. 88).

To sum up, the immediate and progressive differentiation of the germ, starting from the late blastula, appears to us as resulting from two kinds of factors. From the animal to the vegetative pole, from the dorsal to the ventral side, and probably from the surface to its centre, the germ is enveloped in a loose pattern of **differential activities.** They affect the mode of formative movements, and thereby determine the ultimate position of each part in the whole. They also influence the cell physiology, concerning the frequency of division, the permeability, the yolk resorption, the absorption of fluids, the passive or active growth

[1] Cf. the excellent review of Adelmann, 1936.
[2] Raven, 1933; Holtfreter, 1936; Lopaschov, 1937.
[3] Unpublished data. The epiblastic anlage of the pituitary may be absent, reduced, normal, exaggerated according to the relation between the prechordal entoblast and the epiblast.
[4] Holtfreter, 1933. [5] F. E. Lehmann, 1927.

of the cortical film, the ability to contract, etc. What we call the chorda is the median group of cells where mitotic divisions are scarce, the affinity for water great, and the power of extension maximal. The neighbouring muscle cells are the elements which lengthen without vacuolization and soon hasten in employing their yolk. The kidney elements are the still more lateral cells, which are endowed with such differential growth of the surfaces that they enclose a cavity and form hollow buds lengthening under the epiblast and in a caudal direction. An analogous, physiological definition could be proposed, as an approximation, for the coelomic linings, the heart, the blood vessels and corpuscules, the gills, and probably the main entoblastic components. This only leads us to assume that differences in the rate of elementary cell functions are probably responsible for the first steps of organogenesis, inside the marginal zone. These may be called endogenous factors of organogenesis. The second kind of factors is exogenous in the sense that influences originating from the same material act upon adjacent cells by induction, or perhaps, in some cases, mechanically. This is particularly conspicuous in the evolution of the ectoblast, but also seems to affect the entoblast, at least in its anterior and posterior regions.[1] For the reasons already stated, ectoblastic induction appears to be chiefly dependent on the concentration of organizin produced, as a kind of secretion, by the underlying material. Is a correlation to be admitted between the inclination of the invaginating cell groups and their capacity for induction? A direct correspondence may be excluded, for induction is not bound to life nor to progressive organization of the inducer. Moreover, anlagen with a pertinent and rather firm inclination, ventral mesoblast, heart, pronephros, seem, so far as we yet know, to be deprived of inductive activity. But it must be remarked, on the other hand, that the region of the strongest inductive power is also that of preponderant formative movements and of most specialized differentiations: a common condition seems responsible for these three high activities.

Considering the whole process of early development and the results of its analysis, neurulation clearly appears as the automatic

[1] Bytinski-Salz, 1931, for posterior induction; for anterior induction of the branchial pouches, cf. p. 67.

consequence of gastrulation, and further steps seem to be controlled by the same causal factors. While inductions are still working, the maturing individuality of inner organs has its repercussions on the moulding and growth of the ectoblastic derivatives, especially of the brain, the spinal cord, the sense organs; and the outlines of the embryo are definitely traced. Its now typical organs have attained their full individuality, and are not only able to preserve it during the whole life, but also in most artificial conditions of graft and isolation. The real mosaic stage of the prefunctional period is fully established. It is important to state here that the organ- and tissue-specificity which has thus been acquired has most probably a chemical basis. When young gastrular ectoblast, still indifferent, is grafted in advanced neurulae, it develops according to the surrounding local influences, which act upon it through a kind of contagion. The most varied organs, brain, gills, chorda, pronephros, are equally active in this respect.[1] It may be at least suggested that their structural individuality is bound to the possession or constant production of some specific substance, transmissible to adjacent material, provided it still be indifferent. If such a view comes to be substantiated by further researches, it could be of a real importance for the problem of Cancer. The most general aspect of neoplastic proliferations is that tissue cells escape the "individuation field" to which they have been subordinated during development.[2]

Being now aware of the most significant facts concerning causality of development from the end of segmentation, we are prepared to be confronted with the **regulation** observed in experiments made on these stages. The impression of an apparently teleological character given by certain results is caused both by the completeness of the embryo obtained from a partial germ and by the normal proportionality of its organs. In that direction, the discovery of morphogenetic induction, with all its outstanding consequences, has brought about the most substantial progress. The conditions under which a germinal system may exert a normal or abnormal neurogenous induction and consequently give rise to a complete, or partial, or abnormal embryo are now elucidated and, to a certain extent, can be foretold in regard to

[1] Holtfreter, 1933.   [2] Waddington, 1935.

unexplored cases. An excellent instance of recent advances in that field is the interpretation of the *duplicitas cruciata* obtained by a sagittal constriction of a young newt gastrula. The division of the anterior head is combined with the formation of a trunk on the ventral median line. This is due to the acquisition, by the ventral mesoblast, of an inductive power, probably transmitted by contact from the main organization centre. But the presumptive significance of the chordo-mesoblast has not undergone the least change. So far as the middle layer is concerned, regulation has been purely topographical.[1] What is less understood is the capacity of the organization centre to complete itself, to restore its intimate symmetrical structure and to change the fate of some of its parts, apparently with some effort towards "wholeness". In spite of its complex nature this most fascinating capacity of essential regulation does not seem to be out of reach of our actual analysis. In the graft of an organizer, its moulding action on the adjacent material may be assumed to depend on a diffusion of organizin or some similar process. This condition being realized, the qualitative differentiation of the affected material is one and the same problem as the normal individuation of chorda and mesoblast, and as the gradual formation of the somites. If a satisfactory explanation was offered for these typical steps, a puzzling aspect of regulation would be solved.[2]

[1] Kitchin, 1936.

[2] The theory put forward here has recently received support from the confirmatory results obtained by Luther (1937), Yamada (1937), Vogt (1938) and Dalcq (1938). These advances help to elucidate the meaning of Child's physiological gradients in their application to the very early stages of ontogenesis. For a recent complement to Chapter vii see A. Dalcq and J. Pasteels: Potentiel morphogénétique, régulation et "axial gradients" de Child (*Bull. Acad. Méd. Belg.*, 1938, sér. vi, 3, p. 261).

The organization of the Vertebrate egg before and during cleavage (with a theory of early Vertebrate development)

When we attempt to trace the organization recognized in the blastula of a Vertebrate backwards to the newly fertilized egg, we find all our information gathered round the idea of a **primary inducing centre,** and particularly of the conditions preparing its gradual realization. Its precocious existence is demonstrated, in Amphibia, directly by the separation at various stages of blastomeres having a different presumptive significance,[1] and by the complete or partial ligature of unsegmented eggs;[2] indirectly by the cross-superposition of two 2-celled eggs,[3] by the removal of one or two dorsal micromeres,[4] etc.

With other Anamniotes, the situation varies a good deal. In the lamprey, the recent observation of isolated, equal or unequal blastomeres indicates that, in conformity with the anlagen map (fig. 15), the organization centre is exceptionally narrow.[5] It must, on the contrary, be really broad in Teleosts as *Fundulus,* where all parts of the egg exhibit a considerable power of regulation until a relatively late stage of gastrulation.[6] Something similar exists in the trout egg, which was recently studied by grafting pieces of the blastodisc in the yolk sac of an alevin.[7] In the morula and blastula stages, any piece constitutes a complex of all organs and tissues of the three layers; the fact is especially interesting concerning ventral fragments, which have, as shown by fig. 16, a quite limited significance. They are, indeed, when gastrulation begins, the first to restrain their amount of differentiation in those experimental conditions.[8] Gradually, then, the pluripotency

---

[1] G. Ruud, 1925. [2] Spemann, 1928.
[3] Mangold and Seidel, 1927. [4] Votquenne, 1933.
[5] Montalenti and Maccagno, 1935. [6] Hoadley, 1928.
[7] Luther, 1936.
[8] The influence of hormonal or nutritive conditions possibly provided by the host used in such experiments must not be lost sight of. Analogous

becomes restricted to the dorso-marginal organizing zone. It is
to be noticed that the result of isolating, at very early stages, the
ventro-marginal regions of the trout egg calls to mind the behaviour of aggregated Amphibian organizers (p. 63).

Among Amniotes, Birds alone furnish some evidence favourable
to the existence of an organization centre previous to gastrulation.
After a long and diverse series of investigations, the data of which
were differently interpreted by their authors,[1] consistent results
have been arrived at by two distinct methods. Prospecting the
regional potencies of the unincubated chick blastoderm by chorio-
allantoic grafts has revealed that only the dorsal sector yields
all the types of organs, including the neural derivatives; other
regions only differentiate into small pieces of gut, smooth muscle,
heart, liver and skin.[2] The region endowed with totipotency thus
exactly covers the main presumptive anlagen (fig. 23, a). On the
other hand, a methodical study of sections and localized deficiencies in the in situ blastoderm has shown that the most
susceptible region is a median strip extending from the centre
to the dorsal (or posterior) edge. Its total removal generally
reduces morphogenesis to the formation of mesenchyme and
blood vessels, contained between indifferent epithelia. It seems
that the most posterior half of the median strip is especially
important, for its destruction may be followed by a complete
inhibition. With somewhat more anterior lesions, made either
with a knife, or ultra-violet light, or an electrode, three kinds of
embryos may be obtained. In a first group, the embryo is normal,
but extraordinarily small, having developed from the material
located behind the hole caused by the operation (fig. 33, a). In
a second group one embryo is formed on each side of the central
hole, with a complete trunk, having nearly symmetrical somites;
when these twins are equal, their heads are regularly deficient
in the fore-brain; when they are unequal, the bigger member of
the pair may present a normal head (fig. 33, b to f). In a third
group, the embryonic formations are more or less complementary

manifestations of pluripotency have been registered with grafts of young
Amphibian material in the orbital cavity of tadpoles (cf. Bautzmann, 1929).
These results are easily understood from the theoretical point of view
explained in the present chapter.
[1] Cf. Hoadley, 1926, 1927; Kopsch, 1934.
[2] E. Butler, 1935.

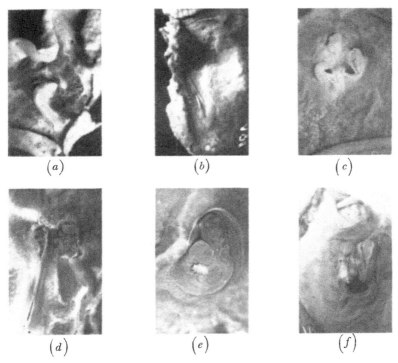

Fig. 33. Results of a sagittal, antero-posterior electrolytic lesion of the chick blastoderm operated before incubation or during the first hours of development. (a) A dwarf embryo, with reduced prosencephalon, obtained by a central lesion of the unincubated blastoderm. (b) Two distinct axial systems, with very reduced heads seen from the vitelline aspect (same operation). (c) Twin, but unequal embryos developed on each side of the lesion made after a 4 hours' incubation. (d) Twin, but unequal embryos obtained by the same operation of the unincubated blastoderm. The heads are both reduced, the trunk of the right individual is incomplete. (e) Same operation. The embryo formed on the right side of the lesion is nearly normal, with a visible otic pit, but its trunk is curved to the left. In the concavity, the secondary embryo is to be seen with only a large neural plate and some somites. Both are enveloped in the amnion. (f) Same operation. The right individual again nearly normal, with its two distinct otic pits; the left embryo much more reduced, with a single, left otic pit, and an asymmetrical troncal region. Unpublished, unretouched photographs of Twiesselmann.

to face p. 78

pieces, or a rudimentary head is separated from the main embryo, or other disjunctions are observed.[1]

These results demonstrate the existence, in the dorso-posterior sector of the chick blastoderm, of a region which is of special importance for morphogenesis. It includes the main presumptive areas; when this region is partially destroyed, a dwarf embryo is obtained; when it is still more reduced, the formation of the embryo is abolished; when it is divided, regulation—but always somewhat incomplete in the head—occurs in a certain number of cases, and in other instances partial structures are observed. All this is entirely comparable to the data concerning the Amphibian organization centre; in spite of the absence of graft experiments, we may consider this centre and the specially endowed area of the chick blastoderm as homologous.[2]

The fundamental unity of the presumptive topography, demonstrated by refined observation, and the unity of morphogenetic functions, discovered by closely studied experiments, are in perfect agreement. This clearly indicates that the **kernel of the morphogenetic problem** lies, for Vertebrates, in the outstanding importance of the dorso-marginal zone, and we have now to examine the origin of this essential feature.

The key of the enigma will be given to us by an attentive analysis of the old experiment of O. Schultze. In 1894, he discovered that a frog egg, when pressed and turned upside down during its early cleavage, undergoes an abnormal gastrulation and often forms a double embryo. The change obviously results from the displacement of certain egg constituents under the effect of gravity. This experiment has been submitted, in these last years, to a methodical and quite productive study. In a Japanese species, *Hynobius lichenatus*, the inverted egg forms its blastoporal lip nearly in its presumptive place, entirely blackened by the ascent

[1] Twiesselmann, 1935–38. The results do not appreciably vary inside the first ten hours of incubation.

[2] It is a striking fact that the three groups of research, chorio-allantoic grafts, deficiencies of the susceptible area and study of the presumptive areas, have been led simultaneously and independently to entirely consistent results. Double embryos after section of the blastoderm have also been obtained by Morita, 1937.

of the pigment, but the direction of invagination is completely reversed.[1] In *Bufo vulgaris formosus*, a well-dosed centrifugalization sometimes causes a slight removal of the black pigment from a region of the animal pole. An invagination appears there, and a secondary tail is subsequently formed.[2] In *Rana fusca*, inversion of the pressed egg is efficient up to the 8-cell stage. The abnormalities of embryonic organization depend on atypical gastrulation. A new blastoporal lip appears in close contact with —or surrounding—any place where some important mass of yolk has fallen, and has been condensed.[3] In many cases, there is undoubtedly a correlation between the original position of the

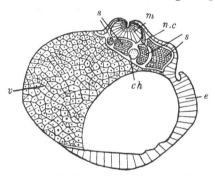

Fig. 34. Superposition of the four micromeres of a frog egg, at the 8-cell stage, on a mass of yolk cells. In such a compound the micromeres, which are unable to develop by themselves, or nearly so, form typical axial organs. The foundation of yolk cells is seen on the left. *ch*, notochord; *e*, ectoblast; *m*, spinal cord; *n.c*, neural crest; *s*, somites; *v*, mass of yolk cells. Redrawn from Vintemberger, 1936.

grey crescent and the place of invagination. This relation is, however, not obligatory. In other inverted germs, the blastoporal lip is initiated only by the accumulation of yolk, and it has been shown that a purely ventral blastomere of the 2-cell stage is able, after inversion, to gastrulate and build up the axial organs.[4] The rôle of yolk in morphogenesis is confirmed by the following fine experiment (fig. 34): the four micromeres of the frog egg (8-cell stage), when isolated, are practically deprived of morphogenetic

[1] Motomura, 1935, p. 218.   [2] Motomura, 1935, p. 223.
[3] Penners and Schleip, 1928; Penners, 1929.
[4] Penners, 1936.

capacity—in spite of their possessing an important amount of presumptive chordo-mesoblastic material; they become able to gastrulate and to form a rudimentary embryo when they are superposed on a mass of purely vegetative cells taken from a very young gastrula.[1]

These facts have induced Pasteels once more to examine, during the last laying seasons, the effect of inverting the eggs of *Rana fusca* and *R. esculenta*. In the first species, the relation between yolk and grey crescent has been especially studied. The inversion was made about the moment of first cleavage, and its degree chosen in such a way that the influence of yolk and crescent would present definite combinations regarding their direction and intensity. The distribution of yolk and black pigment is directly affected by inversion. The slate-grey differentiation of the pigment layer in the grey crescent thereby becomes invisible, but the whole body of results indicates that it may be assumed to be bound to some field arrangement of the superficial pellicle[2] which remains perfectly stable. The topographical and chronological features of gastrulation, in these eggs, are very instructive. In eggs where one principal mass of yolk lies near the centre of the field, which previously could be recognized by the grey crescent, the blastoporal lip appears on the side of the yolk turned towards this centre, and the embryo is normal (fig. 35, *a*, *c*, *f*, *g*). If the yolk mass lies at some distance from the centre, the blastoporal lip may happen to be located symmetrically with regard to the field (fig. 35, *b*) and the embryo is still normal. But if the blastopore happens to have an asymmetrical situation relatively to the centre and to the yolk, the side of the embryo more influenced by the cortical field is distinctly larger than the other (fig. 35, *d*, *e*). In other cases, a small block of condensed yolk has persisted at the primitively vegetative pole; two blastoporal grooves appear simultaneously, and a double embryo is formed (fig. 35, *h*, *i*). If yolk has been agglomerated in two distinct limited masses (fig. 35, *j*), each one will show the formation

---

[1] Vintemberger, 1936. An analogous observation has been made on *Fundulus* eggs. Isolation of the blastodisc from the yolk inhibits any morphogenesis when performed at a sufficiently young stage (Oppenheimer, 1936).

[2] The importance of this pellicle, true limit of the egg, has already been emphasized by Ancel and Vintemberger (1933) in connection with the fertilization reactions.

of a blastopore on the side turned toward the centre, but the one
nearer to it will appear some hours sooner, become larger, and
give rise to a bigger embryo (fig. 35, k). Three types of relation
are thus discovered between the accumulated yolk and the field

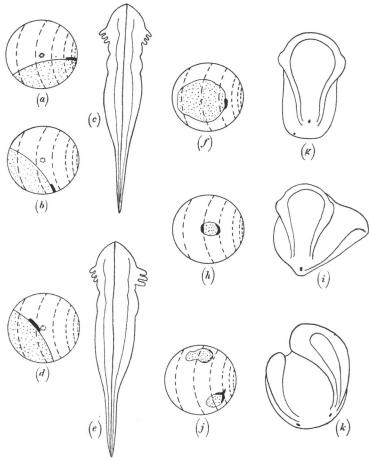

Fig. 35. Schematic representation of the interaction between yolk and the
grey crescent field, as revealed by inversion of the frog egg about the 2-cell
stage. Yolk dotted. Cortical field indicated by concentric circles, in broken
lines. The location of the primitive animal pole is indicated by an asterisk.
Blastopore heavy black. Other explanations are given in the text.

revealed by the grey crescent: topographical, chronological and dimensional. They have also been recognized in *Rana esculenta*, in spite of the fact that the cortical differentiation is at first sight less evident. They govern entirely the initial steps of morphogenesis, the various anomalies of which can be reproduced at will. The experimenter can actually attain the factors which determine the location, the moment of appearance and the symmetrical aspect of the blastoporal invagination. Where this takes place, all the concomitant processes of development also occur. The pattern of regional predispositions makes its appearance and events run their course to determination. The general meaning of these experimental results must now be attentively considered.[1]

Relating the facts described above to normal development, we may conceive the organization centre and its accompanying regional inclinations as dependent on the combination of two factors: the yolk and something normally bound to the grey crescent. The yolk acts by its mass, but also by the orientation of its gradient, for the invagination is always directed from the more vitelline to the more cytoplasmic region. This direction gives that of the cephalo-caudal axis of the embryo, the orientation of which is thus under the dependence of the *yolk gradient*, which shall be designated by V. The factor "grey crescent" is especially rigid. Precise marking shows it to resist the thorough internal change consecutive to inversion, which affects not only the yolk, but also the black pigment of *Rana fusca*. It must consequently be located in the most external pellicle of the egg.[2] Another character of this cortical factor, which will be represented by C, is that its effects concern regions remote from the grey crescent. While no organizing power ever appears in the sub-blastoporal yolk mass, experiment reveals the presence, in that territory, and also at the animal pole, of cortical factors which may enter into a positive interaction with the yolk elements. The cortical factor, C, is thus distributed in a field presenting its maximum in the centre of the grey crescent, at the very place of the future blastoporal lip; from this centre or focus, its intensity decreases. This is the *dorso-ventral field*. Considering the type of reaction occurring in various regions of the yolk relatively to

[1] Dalcq and Pasteels, 1937.
[2] Also assumed by Motomura, 1935, p. 227.

the marked grey crescent, it may be assumed that the decrease of intensity is at first rapid near the focus, then slower in the main part of the egg and insignificant on the ventral side, where nevertheless the field does not fall to zero.

For the first time, we are now confronted with a *concrete representation of the factors promoting the cell to the rank of a germ*. They are not hypothetical entities, but actual data of observation deduced from well-known facts concerning common eggs. The moment has thus arrived to bring back to our minds all the normal and experimental biological reactions of the Vertebrate egg, all the questions that have progressively arisen from our survey of that field of research, and to see whether the interaction of the yolk gradient and of the dorso-ventral field can provide us with the solution of the numerous problems which are raised. To be really satisfactory, that solution must at least fulfil the following requirements: to remain as close as possible to the objective structure of the recently fertilized egg; to reveal the functional sequence from that initial structure to the earliest morphogenetic events, the pattern of regional predispositions and the first changes in the form of the germ; to explain the shape of the presumptive areas, the course of their boundaries and the orientation of the primitive segments; to account for the results of the various experiments, including those of the grafts of organizers, and especially for the appearance of regulation; to give an interpretation of determination; to show the meaning of the power of specific assimilatory induction (p. 75) exhibited by the organs of the embryo.

Let us imagine the Vertebrate egg as an ordinary cell, possessing its normal and complete nucleus but endowed with the yolk gradient and the cortical field which constitute this cell as a real germ.[1] The yolk gradient has been established during oogenesis. It implies, by the stratification of all the inclusions designated as yolk or deutoplasm, differences of density. After the fertilization reaction, this causes the orientation of the egg by gravity, an initial event that cannot be considered as revealing a new factor of morphogenesis, but merely as resulting from the quite passive arrangement of granules laid down in the cytoplasm

[1] A mention must also be made of some plasmatic differentiation preparing the germ for the following generation (cf. p. 102).

during ovarian growth. The gradient varies in the whole mass of the egg, from the vegetative to the animal pole. The decrease begins, however, only at some distance from the vegetative pole, as is evident for telolecithic germs. The minimum attained at the animal pole is still positive, although very low in certain cases, as in Teleosts. The law of decrease is unknown and we shall suppose it to be a linear function between a maximum and a minimum.

The assumption of a cortical dorso-ventral field is indeed inspired by the ideas of P. Weiss,[1] who has so rightly emphasized the usefulness of the field concept for the interpretation of development. The originality of the scheme suggested here is that the field is meant to be located in the very surface film of the egg. This is no gratuitous hypothesis, but a direct deduction of experience. As indicated above, the presumptive dorsal region, recognizable thanks to the grey crescent, still exerts a morphogenetic influence after the complete disturbance of the pigment layer by inversion. Such effect can only be attributed to the very surface film of the egg. *This pellicle must thus be considered, in spite of its extreme thinness, as possessing an important and stable ultramicroscopic structure*, with a decrement of some property from a focus, which characterizes the dorsal side of the germ.

It could be objected that this view is not reconcilable with the modifications of the surface apparently occurring after fertilization and during cleavage. Concerning fertilization, the initial reaction of the egg and the elevation of the membrane may be thought of as a process of permeation which does not necessarily affect the field structure. Secondary remarkable changes are: in the Amphibian egg, the appearance of the fertilization spot(s), and the formation of the spermatic trail(s). There is visibly a flow of pigment towards the place of fertilization, where a centripetal current is formed. Such conditions may be interpreted, by reference to the study of intracellular currents,[2] as resulting from an elevation of surface tension in the region where the sperm has penetrated. This elevation, however, does not affect the interface film-perivitelline liquid, but the interface film-cytoplasm: the granules of pigment actually glide under the

[1] 1930, 1935.   [2] Spek, 1918.

film. Later on, still in Amphibians, other changes appear in the distribution of pigment, and they herald the morphogenetic activity of the germ. This importance of the grey crescent and other homologous features cannot be over-emphasized. They may be considered as the expression of the field properties inherent in the film, which affects, on its inner surface, the cohesion forces of the cortical cytoplasm. When the segmentation occurs, the surface film must of course be somewhat stretched on the interface of the blastomeres. But it has been recently asserted as a result of vital coloration[1] that a real cortical growth takes place in that region and, on the other hand, the separation of the first blastomeres often results in a more or less marked deficiency on the inner side. It seems thus that the morphogenetic element of the film remains rather peripheric, and is lacking in the depth of the furrows.

The focus of the dorso-ventral field is located, relatively to the axis of the yolk gradient, at a variable height in the different zoological groups. In Amphibians, which are always the reference type, it may be placed, according to the situation of the grey crescent and of the future blastoporal lip, about the equator of the recently fertilized egg for Anurans,[2] somewhat lower for Urodeles. In Cyclostomes, its situation is probably the same. In telolecithic Anamniotes, we shall place it at the margin of the blastodisc, while it must be intradiscal in Sauropsides (p. 16). In Amphibians, the densest part of this field is, in some species, rendered visible by a special aspect of the pigmentation, but this seems to be nothing but an epiphenomenon. The active substance is spread on the whole region which is about to be cleaved. Before the segmentation, the C field is purely superficial, varying between a maximum and a minimum. During cell division, it is able to undergo some local extension in the depth along the cleavage furrows.

The data obtained from the inversion experiments allow us to assume that the vitellus and the cortical substance(s) enter into reaction in every point of the egg surface, except where the exaggerated yolk density causes a real inhibition. According to the law of mass action, the products symbolized by C, V are formed in a different amount for each point of the egg surface. This

[1] Schechtman, 1937.   [2] Cf. Votquenne, 1934.

value gives the *morphogenetic potential* (p. 76) and its influence may vary according to (*a*) its absolute amount, (*b*) the proportion in which C and V were originally present at the site of the reaction. The chemical nature of the process must, of course, be left an open question, and the same is true for the influence of an initial excess of C or V, which may affect the first reaction or some further interaction with a pervasive cytoplasmic or nuclear element. Let us now introduce the concept of *threshold*,[1] the use of which is now so general in all quantitatively studied physiological activities. A $C.V$ product exists at each point of the egg surface, and the differences of functional activity expressed by the formative movements indicate that the level of the morphogenetic potential has a decisive importance. The first differentiation (*sensu lato*) to be considered concerns the territories that invaginate and those that remain superficial. This segregation may be attributed to the existence of a definite threshold. The cells which surpass it undergo the general deformation causing their invagination. The formation of the blastoporal lip and its progressive extension simply reveal the region of higher morphogenetic potential: there, the *threshold of invagination* t_i is attained first. The cells whose $C.V$ remains less than t_i perform the simple extension on the surface, which characterizes epiboly.

To make precise these ideas, we will take an instance, based on *Rana fusca*, where the cortical focus F is situated at the equator (fig. 36). Later, by the formation of the blastocoele as a consequence of the cleavage, the same point F, transformed into the focus of the field of morphogenetic potential, will be placed lower, but to avoid complication in the drawing we think it advisable to neglect this displacement. When, starting from F, where $C.V$ is a maximum, we follow the equatorial circle FO, the value of V remains unchanged, while that of C constantly decreases. The invagination will thus begin in F and spread towards the opposite point O: it is the line of the virtual blastopore,[2] which gradually becomes real. A ventral lip will only be formed (Anurans) if the value of $C.V$ attained in O remains

---

[1] This mode of explanation has been at least suggested by Huxley (1935, p. 274): "such threshold phenomena are probably of great importance in development, and are essential to the understanding of epigenesis."

[2] Term introduced by A. Brachet.

superior to the threshold. In certain species (Urodeles) this condition is not satisfied. The material below the virtual blastopore is in a peculiar state because of its excessive charge of yolk and its late segmentation; we may, however, state that it is progressively carried along by the invagination forces. When, starting from F, we follow the sagittal meridian up to the animal pole A, V and C decrease simultaneously. The distance from F, where the invagination threshold will no more be attained, is thus shorter on FA than on FO. Let us suppose that at the point B the decrease of C and V is such that $C \cdot V = t_i$, and let us assume, for the variations of C and V, the arbitrary values respectively represented by the curve (α) and the line (β). Calculating the position of the points of a $C \cdot V$ equal to that of B, we obtain the line BP. This is represented as projected on the median sagittal plane of the egg (plane of the drawing) and its real form on the sphere should be more curved. But this representation is accurate enough to show that the limit of invagination obtained by our hypothesis is conformable to the data of observation: the marginal zone necessarily takes the shape of a girdle, higher in the dorsal than in the ventral region.

Let us now consider secondary segregations, so characteristic of the "Individuation field".[1] The special line BP has been calculated as an equipotential line. Between F and B we may draw, from the intermediate points a, b, c, d, e, other equipotential lines: they have the characteristic obliquity of the somites. But metamerization is not limited to the somites. There is, in the whole embryonic body, a constant relation between the materials of the notochord, epimere, mesomere and hypomere belonging to the same transversal segment. The limits of these general segments are always disposed, on the presumptive map, on a line more or less oblique—according to the species— relatively to the median plane (fig. 24). That direction corresponds to the equipotential lines.

In the territory of the marginal zone, equal areas separated by equipotential lines may be considered. The substance $C \cdot V$

---

[1] This most useful expression has been coined by Waddington: "I wish to suggest the idea that an organization centre is surrounded by a field, which may be called the individuation field, within which its unifying action makes itself felt" (Waddington and Schmidt, 1933, p. 555).

will then be the more concentrated the nearer the area under consideration is to F. It must be remarked, concerning this point, that there is no reason to assume that the reaction $C.V$ only proceeds up to the level of the invagination threshold; its limit depends on the initial concentration of C and V, and this concentration decreases from F to the line BP. Now, it has been established that the capacity of primary induction is confined to a dorsal crescent delimited in the whole marginal zone. If we admit that the intensity of induction is a function of the $C.V$ concentration, we have an immediate explanation for the methodological distinction between head and trunk organizers (p. 59). At a given moment of the development, let us say, at the beginning of gastrulation, the organizin, product or by-product of the $C.V$ reaction, will be far more concentrated in the neighbourhood of F. The objection could be made that the first material to invaginate has no important inductive function; that it goes ahead of the prechordal plate, in the walls of the pharynx, and cannot be made responsible for the induction of the brain (p. 31). But these cells adjacent to F and endowed with the highest potential are scattered and mixed with deep elements (fig. 14), probably as a mechanical consequence of their connection with the underlying entoblastic mass. It is thus to be understood that their inducing power is thereby reduced to the production of the buccal depression. From the presumptive prechordal plate to the limit of invagination, we find a scale of $C.V$ values which will be found again, somewhat modified of course, in the invaginated middle layer. This material in its turn forms a field which may be assumed to act upon the ectoblast according to various induction thresholds. The more important of them affords the distinction between the neural and epiblastic areas of the ectoblast. The epiblast thus results from a positive, but lesser induction, and the exact outline of the neural plate is the projection of an equipotential line running in the substrate. The transition is, however, not abrupt; an intermediate surrounding ridge will represent the particular position of the neural crest, proceeding, in Amphibians[1] at least (fig. 11), both from the lip of the neural tube and from the adjacent epiblast, and typically developing as a link between the skin and the central nervous organs. In

[1] Raven, 1931.

a word, induction depends on the transmission of C. V substances or derivatives of these; they are the organizin, and, contrary to the stable element of the dorso-ventral field C, they are transmissible, through the cortical film, from richer into poorer cells. The inducing elements are not exclusively superficial; on account

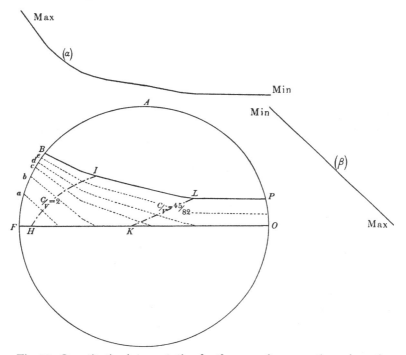

Fig. 36. Quantitative interpretation for the map of presumptive anlagen in an ideal Vertebrate egg. A, animal pole; F, maximum focus of the cortical, dorso-ventral field. (α) Supposed variation of this field intensity from the focus F to the opposite pole O. (β) Supposed variation of the yolk gradient from the animal pole to the equator. Further explanations in text.

of the slight extension of the field in the depth during cleavage, the primary induction centre (organizer) acquires a certain thickness. Materials situated in the dorso-marginal wall of the blastocoele are in fact not deprived of inductive capacity.[1] This quantitative interpretation of induction is thus based on the very

[1] Dalcq, 1936.

conditions which determine the general arrangement of the formative movements. The correlation of both activities has been often noticed, without receiving an adequate interpretation. Now, these dynamic features appear, on the other hand, as the prelude to the forthcoming differentiation of the chordo-mesoblast, another important problem to be faced. The general disposition of the organs in the middle layer immediately suggests that the initial relative amounts of the factors C and V here play the principal rôle. The chordal presumptive area is located in a region where C was generally nearer to its focus than V, while the lateral plate area was more influenced by V than by C, and the somites were in an intermediate situation. If we suppose that a certain level of the ratio C/V provides, by a special modulation of the $C.V$ reaction, the distinction between the three zones, we may give to the chordo-somitic threshold an arbitrary value $C/V = 2$ and to the somitic-hypomere threshold the value[1] $C/V = \frac{4\,5}{8\,2}$. The construction gives the lines HI and KL, and the general disposition represents with a sufficient accuracy the normal repartition of presumptive areas. If certain of the variables, to which we have just attributed arbitrary values, could in the future be measured, the study of the maps would enable the calculation of the real differential thresholds.[2] Here, also, the relative thickness of the potential field must be considered to account for the situation in the depth of the hypomere material in the Anuran egg. The suggested representation easily explains such topographical differences between related species. It also affords a clearer idea of the intimate cause of the somitic segmentation. This process has been attributed (p. 25) to some maturation of substances affecting the surface equilibrium and thereby the form of a group of cells. It actually seems to depend on the gradual attainment of a given $C.V$ value, at first in the group of cells nearer to F, then in a second more lateral group, etc. Thus are the successive waves of somitic formation, which is the fundamental feature of the Vertebrate organism, rendered intelligible by the concept of threshold.[3]

[1] Also quite arbitrary, corresponding to the construction that was drawn by chance.

[2] The boundaries of the organization centre, as inducing material, probably depend on analogous conditions.

[3] Interesting hints concerning the mathematical aspects of morphogenetic

Our comprehension of the formative movements is also materially improved. The most immediate effect of the local morphogenetic potential bears on cell shape. As soon as their size has been sufficiently reduced by karyokinetic divisions, the blastomeres cease to preserve their spherical or slightly polyhedric form. This means that their surface increases, and in different ways for the various territories. Here, some part may perhaps be played by secondary conditions, such as an easier respiration at the animal pole and all around the blastocoele, the free space offered by this cavity, and also the mechanical resistance of the concentrated vegetative yolk. Combined with the local potential, they give to the cell deformation a behaviour characteristic of each territory. According to one specific threshold, the deformation becomes purely extensive, with a slight predominance of the external surface on the internal one, and we obtain epiboly. Elsewhere, the growth of the internal surface predominates, the cells become deformed like flasks, their prominent part bulges toward the blastocoele, and we obtain invagination. Near the medio-dorsal meridian, the growth affects the lateral cell walls more than the animal and vegetative parts of the membranes, and the cephalo-caudal extension is produced. In short, the pattern of the *C.V* field, perhaps combined with some local favouring conditions, imposes on large groups of cells a definite mode of *auxetic surface process*, which is the underlying cause of the formative movements.

By the preceding considerations it has been recognized that the purely *quantitative* interaction of the two specific constituents of the Vertebrate germ cell, the yolk gradient and the cortical dorso-ventral field, suffices to explain the main topographical

problems are to be found in Rashevsky (1933). Assuming two general functions, metabolism and irritability, this author explains the formation of the blastula. Gastrulation is considered as the automatic, mechanical consequence of the given chemical differentiation of the blastula. *Our aim here is precisely to account for this important process.* Later events concerning the neural organization and the segmentation of the mesoblast are also considered in that valuable contribution. The outline presented here seems, however, to be in closer contact with the embryological facts. One of its logical consequences is that *an appropriate flattening of the cortical field will result in the transformation of chorda into somites.* This has been actually attained by intoxication of the blastula by LiCl (cf. p. 63).

and dynamic characters observed in development. Before emphasizing the validity of the theory as an integration of the experimental results, we must come back, for a while, to our **comparative study of the presumptive maps.** There is not much to be added for Anamniotes. The importance of the yolk gradient is clear from the comparison between holoblastic and meroblastic eggs, while the particular case of *Petromyzon* emphasizes the rôle of the cortical field for the localization of the organization centre. The main difference between Anamniotes and Amniotes may also be attributed to a different localization of the same centre. Even if the blastopore of the latter is always intradiscal, it certainly depends on the position of the cortical focus in the cytoplasmic area which undergoes cleavage. The establishment of such a relation between the yolk gradient and the cortical field should be traced, if possible, into the phases of oogenesis. We can only guess that the far-distant mutation which separated the two Phyla affected that essential relation.

This historical process has been connected with other more definite changes. One is the precocious migration of the entoblast, which means, from our theoretical viewpoint, a special production of the *C. V* compounds. This idea agrees with the existence of an intense power of induction in the entoblast of Birds (p. 54). It has been suggested[1] that this particular activity of the Amniote entoblast could be related to the existence of disintegrated yolk in the subgerminal cavity of Reptiles and Birds; a certain amount of this disintegrated yolk is also formed in Mammals.[2]

The other typical change appearing in the development of the Amniotes is precisely the formation of the amnion. It consists, undoubtedly, in an active displacement of the material surrounding the embryonic region. From the end of the last century it has been known that interesting variations are present among Sauropsides regarding the moment when the folds appear, and the direction of their growth.[3] In Reptiles, they seem to be most precocious in lizards and also in the chameleon, where the process

---

[1] Dalcq and Pasteels, 1937.

[2] The most curious polyinvagination described in Birds and Mammals (p. 16) concerns cells charged with yolk or chromophile inclusions, and is thus intelligible from the present theoretical point of view.

[3] Cf. Schauinsland, 1906, in Hertwig, I, p. 177 and Graham Kerr, II, 1919, p. 465.

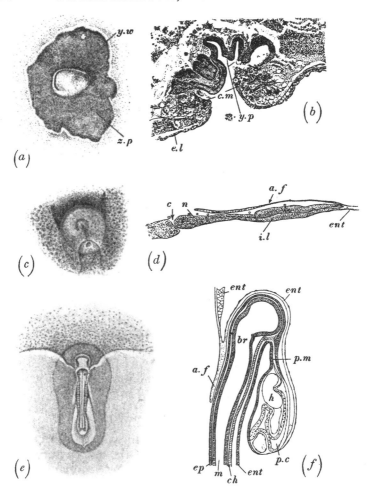

Fig. 37. Some characteristic aspects of the formation of the Amnion.
(a) General view of a chameleon blastoderm, at the moment when the
amniotic folding is formed: z.p, zona pellucida; y.w, yolk wall; the blastoporal plate is just appearing beneath, near the left amniotic fold. (b) The
closure of the enveloping lips, at the vegetative pole, in the same species, to
show the energy of the epibolic tendency: e.l, enveloping layer (ectoblast
underlain by entoblast); c.m, cell masses formed round the cicatrice;
y.p, folded yolk plug. (c) Development of the Amnion in a tortoise (*Clemmys
leprosa*); general view of a young gastrula, the amniotic cavity is still open,

has been accurately studied again in recent years.[1] While the embryonic shield, with its blastoporal plate, becomes more and more conspicuous, the ectoblastic layer stretches over the yolk, closely followed by scattered entoblastic cells. Just at the moment when the envelopment is completed, and the vitelline vegetative orifice closed in an intensely plicated cicatrice (fig. 37, *b*), the amniotic fold suddenly appears all round the embryonic shield, inside the *zona pellucida* (fig. 37, *a*). The closure of the cavity results from a concentric movement towards an amniotic umbilicus, where the amniotic isthmus or sero-amniotic connection is finally formed. In Chelonians, the folding appears somewhat later, and delimits, in front of the medullary plate, a deep cephalic groove (fig. 37, *c*). In that region, no extra-embryonic mesoblast has yet penetrated. The ectoblast adheres to the anterior entoblast, and undergoes a plicature ending with a sharp edge (fig. 37, *d*). The process clearly gives the impression that a powerful extension of the extra-embryonic layers presses against the firmer material of the embryonic shield. Gradually this mechanical folding spreads along the sides of the embryo. The closure of the folds occurs only in the very posterior region, at the extremity of an extraordinarily long amniotic tunnel. Gradually, also, a long amniotic isthmus is built at the roof of the cavity, and the spreading of the extra-embryonic mesoblast separates the entoblast from the external layer. In Birds, the process is not very different, although considerably delayed. Its first indication is seen, in the fowl and in the albatross, when the embryo already possesses about fourteen pairs of somites. At this stage, the enveloping of the yolk is, however, only half completed. The extra-embryonic mesoblast has nevertheless acquired a considerable extension, but is still absent in the

[1] Peter, 1934.

just over the blastopore. (*d*) A sagittal section of the same embryo: *a.f*, amniotic fold, exclusively formed of epiblast; *c*, chordo-mesoblastic canal; *ent*, entoblast; *i.l*, internal layer of the embryonic shield; *n*, neural layer of the same. (*e*) General view of an albatross embryo, with the amnion and the proamniotic region (darkly stippled). (*f*) A sagittal section of the same embryo: *a.f*, amniotic fold, purely epiblastic; *br*, brain; *ch*, notochord; *ent*, entoblast; *ep*, epiblast; *h*, heart; *m*, medulla; *p.c*, pericardial cavity; *p.m*, pharyngial membrane. [(*a*) and (*b*) redrawn from Peter, 1934; (*c*) and (*d*) drawings of Pasteels; (*e*) and (*f*) redrawn from Schauinsland, 1906.]

anterior region, called the proamnion. This is the site of the
first amniotic plicature (fig. 37, *e*), which again is simply formed
of ectoblast and entoblast and seems capable of the same interpretation as for Chelonians. The closure of the amniotic umbilicus
is also remarkably polarized, although not to the same degree,
in the same antero-posterior direction.

It is an extraordinary fact that nobody, among the attentive
students of the amnion formation, has ever noticed the evident
correlation of this process with gastrulation. The sole adequate
remark which has been made seems to be[1] that the earlier the
process, the more the ectoblastic part of the amnion is developed.
Actually—and it is characteristic of the progress introduced by
the kinetic conception of early development—the relation with
the movements of gastrulation cannot escape our attention. The
beginning of the change can be clearly localized. A participation
of the mesoblast is certainly excluded. The folding region is, in
all cases, only formed by ectoblast and entoblast. But this last
layer is not necessarily present, and the disposition encountered
in *Clemmys* (fig. 37, *d*) clearly shows that primarily the ectoblast
alone is involved. There is no special abundance of mitoses to be
observed, only an accumulation of cells, in the folding region, at
the very first stage. Again, in the fowl, a vitally stained mark
placed at the anterior part of the proamnion undergoes a most
characteristic elongation. The whole event cannot, consequently,
be considered as a process of growth, as is still done in classical
descriptions. Obviously, the formation of the amnion is a
secondary and most remarkable result of the ectoblastic epiboly.
The strength of this stretching tendency is evidenced by the
cicatrice of the yolk orifice in the chameleon (fig. 37, *b*). Of
course, the amniotic folding often appears before the end of the
envelopment, but it may be considered that the ectoblast creeping
over the yolk encounters, in the equatorial region, some resistance.
The adherence being there considerable, the powerful epiblastic
stretching causes a folding in the more plastic material of the
periembryonic region, especially in the territory where mesoblast
is still lacking. The similarity between the closure of the cavity
and the dorsal polarization of the gastrular movements is thereby
perfectly elucidated.[2] If such an interpretation is admitted, other

---

[1] Schauinsland, *loc. cit.*   [2] Dalcq, 1937.

consequences appear. The embryo being enclosed in the liquid of the amniotic cavity, a fetal respiratory organ becomes indispensable. The same cause which underlies the origin of the amnion must consequently, to insure a viable germ, have provided the formation of the allantoic diverticle of the enteron. This may have its actual cause in an induction by the mesoblast condensed in the posterior region of the amnion, under the amniotic isthmus.[1]

Some well-known facts concerning placentary Mammals suggest a last conjecture about the meaning of the trophoblast and its villosities. It seems obvious that eutherian Mammals are descended from ancestors already endowed with foetal membranes analogous to those of Sauropsides. This being admitted, some prolongation of intra-uterine life has become compatible with the disappearance of the shell, transitory stages being represented by some viviparous Reptiles. The blood irrigation of the amniotic region by the vessels of the allanto-mesoblast introduces then the immediate utilization of the maternal nutritive supply, a progress which makes it possible to obviate yolk formation. But this achievement does not imply any change in the primary and so deeply ingrained forces of gastrulation. The epibolic tendency of the ectoblast has been preserved, but again with a new utilization: the abundant formation of the trophoblastic villosities. This precedes the appearance of the amnion, just as is the case for the spreading of the extra-embryonic ectoblast on the yolk of a Sauropside egg. The multiplication of the trophoblastic surface by innumerable small processes will be soon combined with the existence of the amnion—which has, in many Orders, undergone certain modifications—and of the allantois, to insure the formation of the placenta.

The **experimental results** that we attempted to summarize concerning the Vertebrates clearly fall under two heads. Some exhibit steps of the evolution towards the final determination, others apparently lie outside the normal line of development. For both, the conception now arrived at offers a satisfactory explanation. It is suggested that, soon after fertilization, a definite topographical and dynamic relation is established between

[1] Entoblastic inductions have been recorded, p. 74

the yolk gradient and the dorso-ventral field. While cleavage proceeds, a progressive reaction takes place between these substances and a resulting $C.V$ field is established. The reaction products that comprise this field are not firmly bound to the cells. Their concentration is responsible for the position of the limit of invagination and for the intensity of induction. If a small territory is grafted in a region of higher potential, its morphogenetic level is accordingly elevated; in the opposite case, it is lowered; in both instances, the fate is conditioned by environment (*ortsgemäss* evolution). If the territory is large, its assimilation is not possible (*herkunftsgemäss* evolution), and in cases where the potential of the graft is high, a new field of potentials is developed, and a secondary embryo is produced. This process, the discovery of which has been the leaven of modern causal embryology, is based on the diffusion of active substances. While, in the normal egg pattern, local differences are too gradual to cause such a diffusion, this happens in the grafts, where regions of high and low concentration are juxtaposed. It results in an addition to the local morphogenetic substances, and in the raising of their concentration above the various thresholds. *The part played by the host germ* is thus intelligible. It is further quite remarkable that the changes involved include the gradual fixing of the already prepared inclinations inside the grafted piece and the formation—at least for sufficiently early operations, performed on the blastula or young gastrula—of a new, mostly symmetrical pattern extending over both the graft and the adjacent host material. This seems to mean that the new field formed with the graft—the potentials of which have become practically homogeneous—as a central region creates by itself the differential quantitative conditions for the distinction between chorda, somites, coelomic linings, etc., i.e. that it becomes an individuation field. The same remark could be made concerning the experiments where a lateral elevation of temperature, at the same stages, provokes, by an acceleration of the chemical processes, the formation of a more or less distinct morphogenetic field, with all its consequences. The significance of such facts is that the separate consideration of the products $C.V$ and of the ratio C/V has only a methodological value, as a basis for a quantitative estimation. In the reality of egg metabolism, both conditions are combined to give a special and gradually diverging orièntation to the reactions going on from the very moment of fertilization.[1] At the blastula stage, the substances characterizing the morphogenetic potential proceed, both in their already complex nature and in their concentration, from the initial value of C and V. Their pattern is easily disturbed by conditions causing a local difference of potential, but is not greatly altered in certain experiments in which, as in translocations, *the position of certain parts relatively to the co-ordinates of the egg system is not appreciably changed*. The theory here presented also improves our comprehension of certain old and recent experiments made on the first stages of cleavage. In several Anurans, punctures (with a cold needle) of the undivided or slightly cleaved egg provoke, in a large proportion of cases, local deficiencies corresponding to the operation; a puncture of the presumptive marginal zone leads to reduction of the somites, and with a lesion of the neural area, a unilateral reduction of the brain or spinal cord has, without any doubt, been observed in some embryos.[2] This experiment simply produces defects in the cortical field that are not always compensated by the mobility of the C. V products. On the other hand, it has been known for a long time that a left or right blastomere of the 2-cell stage, in *Rana fusca*, develops, when isolated, into a hemi-embryo.[3] This means that the surface of separation remains, in the main, deprived of morphogenetic potential. The different reaction of the newt egg, in the same experiment, is perhaps to be related to the difference in the extent of the region capable of induction, which is broader in the Urodele than in the Anuran. The degree of mobility of the C. V products, at the end or during cleavage, may also be of importance.

A few words will be sufficient concerning **determination**. The morphogenetic elements of the egg and its blastomeres cannot be thought of as remaining independent of the cell metabolism.

[1] A considerable change in the respiratory metabolism of the egg has been demonstrated to be a striking consequence of Amphibian gastrulation (see note 1, p. 23). This is quite consonant with the theory now arrived at.

[2] Pasteels, 1932.

[3] It has been verified that the elimination of the debris of the killed blastomere does not change the result (Vintemberger, 1929).

100   ORGANIZATION, CLEAVAGE AND

They, or their derivatives, enter into reaction with other substances, which may be either cytoplasmic or nuclear. About the former nothing very definite is known, except the probable influence of respiratory enzymes. For the latter, a clue is given by the experiments where the gametes are submitted to irradiation or intoxication,[1] with, as a result, a considerable inhibition or even a suppression of the formative movements. Similar facts are gathered from hybridization and are especially instructive when merogony is combined with hybridization (hybrid *androgenesis*); embryonic development is then observed to be inhibited in certain regions, by a change in which nuclear trouble is evident, while healthy parts of the embryo may be safe if grafted at the appropriate moment on to normal embryos.[2] These various facts point to a linkage between the nucleus and the morphogenetic substances; nothing indicates that genes are here involved; other, less specialized constituents of the nucleus are probably responsible for these first steps of its participation to morphogenesis. But genes are likely to play an active part in the forthcoming cyto-differentiation, which is responsible for the last refinements of organization, with all its specific and racial hereditary characters. Two facts give us an idea about these internal and ultimate processes of determination. The embryonic organs are still able to induce indifferent material and bestow on it their special form and differentiation. This means that these properties have a chemical basis represented by compounds which can still be transferred to receptive young cells. On the other hand, the culture of the already specialized cells shows their transformation and dedifferentiation only in a few exceptional cases. The chemo-differentiation is thus intimately and often definitively bound with the whole cell metabolism.

To **sum up**, a fairly consistent representation of Vertebrate morphogenesis can be obtained on the basis of the ideas of gradient, field and threshold. The combination of the yolk gradient, extended to the whole cytoplasm, and of the cortical field, limited to the surface film, produces the secondary, likewise

[1] O., G., P. Hertwig; cf. Dalcq and Simon, 1932, and Dalcq, 1935.
[2] Baltzer, 1930, 1935; cf. Hadorn, 1932, 1937.

cortical field of morphogenetic potentials. To this fundamental pattern, regional fields are progressively superimposed, as was suggested by P. Weiss, by a gradual interaction with internal cell components. Determination is achieved, by chemo-differentiation, when the chemical linkage of the morphogenetic substances has become irreversible. In every case, including the apparently teleological facts of regulation, the development of a part or of a system automatically results from the primary gradient and field, or their immediate derivatives.

A last question arises: **Does the dorso-ventral field exist before fertilization?** Amphibian eggs again provide the only available evidence. In Urodeles, there is no relation between the point of entrance of the fertilizing spermatozoon and the plane of bilateral symmetry. In Anurans, such a relation indeed exists, especially in *Rana fusca*, but it was a long and tenacious error to suppose that the point of fertilization determines the plane of bilateral symmetry. A statistical study shows that the constancy of the relation is not absolute in normal eggs. In experimental dispermy, the symmetry is not, as has been pretended, the resultant of the two excitations. In exactly localized fertilization, no relation exists[1] between the point of entry and the grey crescent, in spite of the credit granted to Roux's experiment, which should now pass under the veil of oblivion. There remains no doubt that the normal condition results from a certain meridian being more favourable to the entry of the spermatozoon. The unfertilized egg should consequently be considered as being endowed with some dorso-ventral organization, which furthermore becomes visible after any kind of activation.[2] In the *Discoglossus pictus* the spermatozoon penetrates so near to the maturation spindle that its point of entrance cannot be thought to have a special relation to one of the meridians and to determine the bilateral symmetry. Finally, *Rana esculenta* furnishes the unmistakable proof of the pre-existence of the dorso-ventral field. It has long been known that the fertilized egg seems to stay oblique in its membrane.[3] Its animal cap of brown pigment reaches a lower level, relatively to the primordial axis, on one side than on the other. The higher

[1] Ti Chow Tung, 1933.   [2] A. Brachet, 1911.   [3] W. Roux, 1883.

region is homologous with the grey crescent. Now, the obliquity of the pigment cap already exists, although at a variable degree, before fertilization. A statistical study shows that its higher edge indicates, in 95 per cent. of 120 cases,[1] the site of the light crescent and of the forthcoming blastoporal lip. The conviction is strengthened by the observation of trails of pigment, descending towards the vegetative pole on the ventral side of the egg, about two hours after fertilization, and presaging the dorsal appearance of the light crescent. If the lower part of the cap is observed before and after fertilization, it is regularly seen to become the site of these typically ventral pigment trails. Again, the point of penetration of the spermatozoon, which is the origin of the pigmented path, has absolutely no relation with the bilateral symmetry. The series of dispositions observed in Amphibians evidently allows the conclusion that the dorso-ventral field exists prior to fertilization, and is not modified by it. But although the topographical relation is not altered, the functional situation, of course, undergoes a thorough change. The inertia of the unfertilized egg means that an inhibition is opposed to the interaction of the yolk gradient and of the dorso-ventral field, both already represented by their special substances. The blockage may be due to some chemical barrier or to the state of the substances themselves. For reasons to be indicated later (p. 182) I feel that it is most probable that the cortical substance is represented, before fertilization, by a precursor compound responsible for the inhibition of the whole cell system. The cortical reaction of activation modifies this precursor, liberates the true cortical factor and possibly the vegetative factor responsible for the evolution of the *germen*. So would be linked together, as will be explained more clearly in the last chapter, maturation, fertilization, morphogenesis and reproduction. My deep conviction is, indeed, that the main processes of development are, in their essence, intimately linked together.

---

[1] 100 per cent. if the eggs with specially evident obliquity are chosen (88 cases).

CHAPTER VIII

The organization of the Prochordate germ, with special reference to the unfertilized Ascidian egg

For the eggs of Vertebrates, our backwardly progressing analysis had to stop at fertilization. Except for some scattered evidence, experiments have so far failed to unravel the organization of the unfertilized egg. This can be done for one Prochordate at least, and we must return to the *Amphioxus* and Ascidian egg. No satisfactory information is available for stages beyond cleavage, but the **potencies of the early blastomeres** are fairly well known in both species.

In *Amphioxus*, when the left and right halves of a 2-cell stage are separated or displaced, or when the same is done for the "anterior" and "posterior" halves of a 4-cell stage, the development of those isolated parts depends on the presence of the mesoblastic and chordo-neural materials[1] (p. 38). Both being present, regulation occurs. This result could be easily understood according to the differential composition of the cytoplasm, were it not for the asymmetry which is produced in the anterior entoblast: both twins present the diverticles with the normal disposition; in neither is the asymmetry reversed. No satisfactory explanation has been offered concerning this asymmetrical development.

One of the earliest embryological experiments established that, in *Ascidiella*, one of the two first blastomeres is only capable of forming a half-embryo.[2] The same difference thus exists as between frog and newt. The same error has also been persistently committed, of interpreting the production of the half-embryos as evidence of a precocious determination. It has been sufficient to use, on Ascidian eggs, methods inspired by what had been done on Vertebrates and Echinoderms to obtain the most interesting series of results. Only two or three stages have, so far,

[1] Conklin, 1933. [2] Chabry, 1887.

been found accessible to operation. But the smallness of the *Ascidiella* embryo and the most precise cyto-differentiation of its organs (fig. 38) permit a thorough quantitative and qualitative study. Recent results of vital staining observations have afforded the most valuable basis for the interpretation of the experiments. The presumptive topography of the 8-cell stage has already been described (fig. 18). The four micromeres and the four macromeres could be isolated and submitted to certain combinations.[1] The four micromeres, the presumptive fate of which is exclusively ectoblastic, are able to gastrulate and form epiblast, entoblast and one or more neural brain-like vesicles (fig. 39, *a* and *b*); exceptionally, chorda and muscle also appear. The normal fate of the macromeres is to form entoblast, chorda, muscle and mesenchyme. When isolated, the group of the macromeres gastrulates, but with a marked deficiency of the ectoblast. Chorda and mesoblast appear constantly, while the formation of a neural plate is occasional (fig. 39, *a* and *e*). Rotation of the micromeres on the macromeres does not produce an obstacle to an apparently normal gastrulation. An elongated embryo is formed; entoblast, epiblast, chorda and muscle are normal, but the neural organ is found at the root of the tail, generally as an irregular vesicle. The sensory cells, when differentiated, are superficial and posterior to the brain, not far from the tip of the tail (fig. 39, *e*). This third experiment clearly indicates that a neurogenic condition exists in the dorsal side of the micromeres. The first two experiments are remarkable for the plasticity revealed in the Ascidian egg in the 8-cell stage. A blastopore is able to appear either much higher or somewhat lower than normally. While chorda and mesoblast are relatively stable, but certainly not rigid, the neural organ is both more susceptible and more plastic. It must also be mentioned, as a result of the rotation experiment, that the adhesive papillae still appear on the anterior region of the body. This means that they are dependent on a condition which is external to the epiblast, and which can only be the anterior entoblast: this is the first case of induction discovered in a Prochordate.

This year, a thorough reinvestigation of the potencies of the 8 and especially 16-cell stage has been made by Vandebroek[2].

[1] Ti Chow Tung, 1934.   [2] Unpublished results.

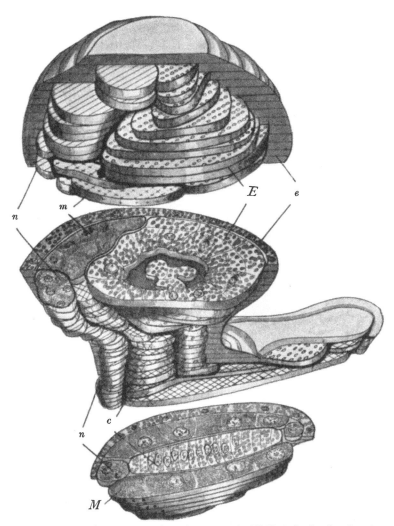

Fig. 38. A graphic reconstruction of a young *Ascidiella* tadpole, showing the arrangement of the organs and their cytological characteristics. *c*, chorda; *e*, epiblast; *E*, entoblast; *M*, muscles or myoblastic cells; *m*, mesenchyme; *n*, neural organ.

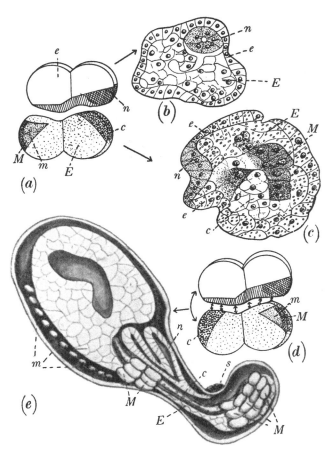

Fig. 39. Experiments on the 8-cell stage of *Ascidiella scabra*. (*a*) Presumptive map, with the separated micromeres and macromeres. (*b*) A section in the structure developed from four isolated micromeres. (*c*) Idem for the macromeres. (*d*) Egg of which the micromeres have been turned 180° round the axis; disposition of the anlage as resulting from the rotation. (*e*) Schematic reconstruction of one of the embryos obtained, especially to show the form and position of the neural organ and of the pigmented sensorial cell. *c*, chorda; *E*, entoblast; *e*, epiblast; *M*, muscular cells; *m*, mesenchyme; *n*, neural organ; *s*, sensorial cell.

This experimenter has succeeded in dividing the germ from its dorsal to its ventral side into four slices. He has been able in this way to produce various combinations of the materials distributed along the dorso-ventral axis. One at least of the interesting results he has gathered by his method may be mentioned here, as the *in vivo* observations are, on this point, perfectly conclusive. The sensorial pigmented cells of the brain are dependent on the influence of a definite part of the entoblast. Counting the four slices in the dorso-ventral direction, it is the third, next to the mesoblastic crescent, which contains the entoblast which comes, in normal development, to lie just ventral to the brain. The presence of this material is necessary for the formation of the so-called eye and otocyst. Conversely, only the brain, and not the truncal part of the neural organ, is able to react to the induction exerted by this prechordal part of the entoblast. As mentioned just above, another part of the entoblast induces in the epiblast the adhesive papillae. The existence of induction processes may consequently be extended to Prochordates, but, so far as we actually know, they only occur in a relatively late phase of development, to cause secondary differentiations of ectoblastic derivations.

For the **fertilized but still unsegmented egg**, our information is rather scarce, as most operative procedures interfere with the movements of pronuclei and consequently with cleavage. On the egg of *Ciona*, Reverberi succeeded recently in performing a series of sections, latitudinal and more or less meridional, at the earliest possible stage. The comparative study of cleavage in the various cases shows that the condition governing the typically symmetrical cleavage is located somewhere in the vegetative part of the egg, distinctly under the equator. Pieces that lack this material cleave atypically; again, the larvae they form tend to be spherical, while those developing from the other type of fragments approach somewhat more the normal conformation.[1]

Let us now turn to the **unfertilized egg** and examine first its **normal structure** and the indications furnished by its vital staining. The egg of *Ascidiella scabra*, easily liberated from its

[1] Reverberi, 1937.

membranes, is a drop of semifluid cytoplasm containing the first maturation spindle. This spindle is anchored to the cortex at a place which is recognizable *in vivo* under the dissecting binocular. Cytological examination of the fixed and serially sectioned egg shows that small eosinophile granules, bearing a pigment which is yellow in the living state, are scattered in the internal ground substance between the yolk platelets. They are more condensed in the superficial layer and cover the whole egg, except for a region surrounding the maturation spot. The part of the animal region where the surface film is made of hyaline plasma is elongated and excentric relatively to the spindle and has thus an irregular oval shape. From the cytological viewpoint, the ripe unfertilized egg of *Ascidiella* has thus a bilateral symmetrical structure, which is, unhappily, not to be perceived *in vivo*.

If one or two vital marks are conveniently placed and the egg fertilized, the coloured spots may be observed until the 8-cell stage, which is already sufficient to appreciate their significance; sometimes they can be followed into the well-formed embryo.[1] All organs can be coloured in this way, except the mesoblast, for which material a coloration becomes only possible when, as a result of the cytoplasmic currents following fertilization, it has been condensed into the yellow crescent; the staining pipette must be left in place a fairly long time, without encroaching on the adjacent epiblast and entoblast, since these materials would otherwise be damaged by overstaining. All this can be easily linked with the data of pure observation[2] and of cytological examination. The cells forming muscle and mesenchyme are those containing the plasm carrying the yellow pigment. Before fertilization, this plasm is located in the above-described thin cortical layer, crowded with eosinophile granules. When attempts are made to stain it, they only result in the coloration of yolk platelets and of other granules contained in the underlying material.

This negative result of the vital staining method being elucidated, its positive achievement is the localization of the subcortical territories belonging to the brain and spinal cord, the epiblast and entoblast (fig. 40). Their extent does not differ from what it

[1] Vandebroek, 1937.
[2] Particularly those of Conklin in *Styela*.

will be at the 8-cell stage (fig. 18); only the neural area is a more regular crescent, with its long and sharp horns inclined towards the vegetative pole. But this similarity does not exclude important changes occurring during the interval between these two stages. The penetration of the spermatozoon, which always takes place near the vegetative pole, seems to cause a lowering of the

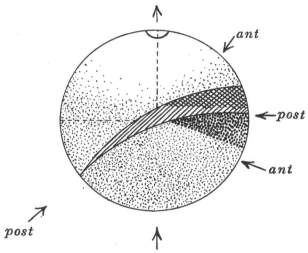

Fig. 40. Map of the presumptive areas in the unfertilized egg of *Ascidiella scabra* and *A. aspersa*. Hatched: the neural organ, with the brain territory cross-hatched; dotted: the chordal area; delicately stippled: the mesoplasm which is supposed to underlie the other anlagen. The limit between epiblast and entoblast lies about the equator. The chordal and neural anlagen partly cover the entoblast. Original drawing of G. Vandebroek (unpubl.).

surface tension at the opposite region. Vital staining of the whole egg enables us to observe an immediate flowing of the superficial layer from the animal towards the vegetative region. At the animal pole, the central point of the centrifugal motion does not necessarily coincide with the exact spot where the polar bodies will soon be extruded. Between the polar spot and the real animal pole, the distance may attain 30° to 35° in any direction. At the vegetative pole, the fountain-like movement causes at first a penetration of entoblastic material inside the cytoplasm; it rises along the axis and will be located finally in the walls of the enteron. The place left at the vegetative pole is taken by the yellow plasm condensed in the well-known yellow cap. The same fountain-like movement produces an important descent of the neural and the chordal areas; but the deeper part of these anlagen is not carried downwards at the same degree, and the neural and chordal material remain so to say anchored to the central region. In the next period, during the conjugation of the pronuclei and probably in relation with the developing asters, the yellow plasm moves along the ventral surface of the egg and forms there the rather thick mesoblastic crescent, while the neural and chordal areas undergo a similar ascension on the dorsal region. In *Styela*, they soon become visible there as a light-grey crescent.[1] The hyaline substance of the surface film, formerly concentrated towards the vegetative region, has gradually returned to a more even distribution.

This exact knowledge of the displacements preparing the formative movements in the Ascidian egg will be most useful for interpreting the **experiments made on the unfertilized Ascidian egg.** An important preliminary remark is that, from the beginning,. a difference exists between the materials forming the muscle and mesenchyme and those building up the other organs. The former are endowed with a sort of primary cytological autonomy; they represent a *mesoplasm*; we have no reason for considering it as determined, but we are aware that a specialized structure is present. For the other organs, there is no perceptible specialization; we only know, by colouring the cytoplasmic inclusions, that such and such a zone will be included in chorda, neural organ, entoblast or epiblast; our information is thus merely topographic.

When an unfertilized egg of *Ascidiella scabra* is exposed on a bed of agar, it may be gently divided with a thin glass thread into two pieces, and both fragments can normally be fertilized. By this operation the surface is evidently increased, and this modification certainly differs from the growth of surface normally occurring in the cleavage of the egg. The extraordinarily delicate

[1] I want to thank again Dr Vandebroek for his full explanations and our friendly discussions about his observations.

surface film—probably of lipoprotidic nature—must here be passively stretched to cover the region of the section, for any direct exposure of the internal cytoplasm to sea-water causes an instantaneous cytolysis. When the operation is successful, a cicatricial region is sometimes to be recognized in the embryos, by reason of its poor differentiation. Cytological examination of recently operated eggs shows that the subcortical layer, crowded with the yellow pigment, is almost unaffected by the stretching of the surface film; the layer of eosinophil granules undergoes an abrupt interruption where the cut has been performed.

The maturation pole being visible *in vivo*, its axis may be used for the orientation of the section. The above-mentioned, unforeseeable discrepancy between the polar spot and the animal pole may cause the intended latitudinal or meridian cut to be, in some cases, slightly oblique, but this deviation is not important enough to hinder analysis. The results clearly fall into three groups according to the direction of the section: latitudinal, meridian and parameridian, i.e. passing through a small circle parallel to the egg axis.[1] Previous vital staining of two or three areas would of course have been desirable; this could not, however, be carried out satisfactorily owing to the technical difficulty of the experiment. But nearly all embryos have been submitted to a thorough cytological study, in serial sections; for the characteristic pairs of embryos, all sections have been drawn with the *camera lucida*, the surface of the organs measured with a planimeter, and separate or combined reconstructions made by a graphic method[2] that makes particularly evident the form and size of the organs (fig. 41).

A first general result[3] of these experiments is that any part of the egg is able to gastrulate. It is not excluded that certain very small fragments may limit their activity to segmentation. But in fragments whose diameter is greater than about $25\,\mu$—that of the egg is about $160\,\mu$—gastrulation is constant. The proportions of

[1] Dalcq, 1932, 1935, 1937.
[2] Lison, 1936.
[3] Most of the fragments divide typically and show, during cleavage, the characteristic bilateral symmetry. This feature may be absent in certain pieces. In *Ciona*, an *extraovate* obtained by puncture of the unfertilized egg divides almost typically (Reverberi, 1936).

ectoblast and entoblast are capable of large variations, especially demonstrated by latitudinal sections.

The structure of the embryos arising from the first type of operation, i.e. *latitudinal section*, varies according to the level of the section (fig. 42). When the cut is made between the polar spot and the equator, the bigger, haploid embryo is normal or

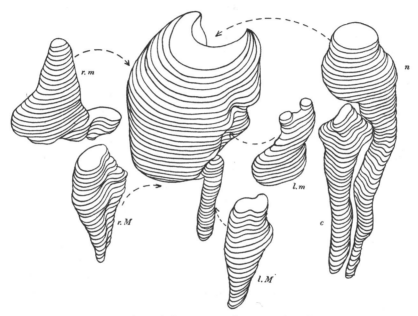

Fig. 41. Reconstruction of the organs of a normal embryo of *Ascidiella scabra* in separate pieces. The embryo is seen from the ventral and left side. Abbreviations as in fig. 38. *l* and *r*, left and right. The stage is slightly older than in fig. 38.

nearly so, but the smaller, diploid one may present different structures, depending on the exact level and probably also on the exact orientation relatively to the animal-vegetative axis of the operated egg. The simplest case, but a rather exceptional one, is an ampulla of ectoblast containing a mass of entoblast, without any other differentiation; the twin embryo is normal (fig. 42, I). In a second instance, a third organ is found between ectoblast and entoblast; its position near the blastopore, the granulations

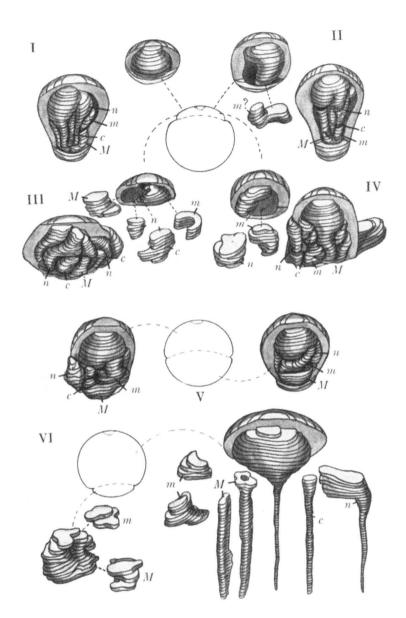

Fig. 42. Results of latitudinal sections of the unfertilized egg of *Ascidiella scabra* (experiments of double merogony). I, II, III, IV, Four possible results of super-equatorial sections. In each case, the two embryos are represented after excision of a large epiblastic region; some organs are separated. V, Equatorial section. VI, Low sub-equatorial section; all the organs of the larvae are represented separately. Abbreviations as in fig. 39.

of its large cells are typical of mesoblast, mostly myoblastic (fig. 42, II). The twin embryo is complete and the proportions of its organs are, as appear from Table I, fairly normal.[1]

Table I

Embryo	Total size (%)	Ento-blast (%)	Epi-blast (%)	Neural organ (%)	Chorda (%)	Mesen-chyme (%)	Muscle (%)	Spaces (%)
An	41	23	53	—	—	?	13·5	10
Vg	59	27	37·5	5·5	6·5	12	9·5	3
Both	100	25·5	44	3	6	7	11	3
Controls	100	±24·5	±35·5	±7·5	±6	±8·5	±10	±8

If the sum of *An* and *Vg* is considered, it appears that the amount of epiblast and of neural organ is very different from what the unoperated egg would have formed. A similar case has been found, where the third organ of the animal part was a distinct hollow neural ampulla, although incompletely differentiated.

In a third operation, the closed vesicle of epiblast contains three distinct masses of entoblast, mesenchyme, neural organ. The larger, haploid embryo is slightly distorted, but nevertheless contains all the organs (fig. 42, IV). The size of these is given, for both larvae, in Table II.

Table II

Embryo	Total size (%)	Ento-blast (%)	Epi-blast (%)	Neural organ (%)	Chorda (%)	Mesen-chyme (%)	Muscle (%)	Spaces (%)
An	18	9	62·5	9·5	—	9·5	—	9·5
Vg	82	25·5	35·5	8	5	15·5	7	3·5
Both	100	22·5	40·5	8	4	14·5	5·5	5
Controls	100	±24·5	±35·5	±7·5	±6	±8·5	±10	±8

[1] In the tables given in this chapter, the two first ranks give the relative sizes of the organs in each embryo; the third, the total volume of each organ formed by both embryos considered together; the fourth, the mean number furnished by comparable controls. The results are given in round figures. A preliminary investigation has established the amount of fluctuation exhibited by the measurement of the organs in the controls.

It shows an augmentation of mesenchyme in the larger embryo (*Vg*). For the sum of *An* and *Vg*, there is, regarding the amount of material formed, an increase of mesenchyme and a reduction of chorda.

Of the two larvae obtained by another section made about the same level (fig. 42, III), the small, diploid embryo is qualitatively complete but the muscular and mesenchymatous cells are not ordered symmetrically, as they should be. The cyto-differentiation of all organs is incomplete and clearly delayed, compared with that encountered in the main embryo, which is, in this particular case, remarkably abnormal, having two distinct neural organs and two separate masses of notochordal cells. The size of the organs of both larvae is given in Table III.

Table III

Embryo	Total size (%)	Ento-blast (%)	Epi-blast (%)	Neural organ (%)	Chorda (%)	Mesen-chyme (%)	Muscle (%)	Spaces (%)
An	19	13	45·5	4·5	18·5	9·5	4·5	6
Vg	81	41·5	24	8·5	4	5	3	8
Both	100	36	32·5	7·5	6·5	5·5	3	9
Controls	100	± 24·5	± 35·5	± 7·5	± 6	± 8	± 10	± 8

In *An*, the reduction of entoblast relatively to epiblast is remarkable, as is also the disproportion between the neural organ and the chorda. The vegetative part (*Vg*) has undergone a considerable reduction of muscles. The sum of the two embryos also indicates that muscles are reduced and entoblast exaggerated.

It is evidently sufficient to draw a somewhat lower cut to obtain, from the animal part, a dwarf embryo of entirely normal constitution. The vegetative part will also build up a normal tadpole, in so far as its gastrulation proceeds satisfactorily. This result is still obtained with equatorial sections (fig. 42, V); there is however a tendency, in the vegetative part, to form a reduced ectoblast, which does not succeed in covering the bulge of the inner materials, and the hind-part of the body will then be abnormal. In the best cases (fig. 42, V) both embryos are normal except for a shortening of the tail. The sizes of their organs are compared in Table IV.

Table IV

Embryo	Total size (%)	Ento-blast (%)	Epi-blast (%)	Neural organ (%)	Chorda (%)	Mesen-chyme (%)	Muscle (%)	Spaces (%)
An	42·5	21·5	39·5	9	9	6·5	10	4·5
Vg	57·5	22	34	10·5	9	11	11	3
Both	100	21·5	36·5	10	9	9	10·5	3·5
Controls	100	24·5	35·5	7·5	6	8	10·5	8

They agree remarkably with the figures of the control. The sole point to be noted is the slight increase of epiblast in the *An* embryo. Two operations of this kind have resulted in a remarkable dissociation, the animal embryo possessing neural organ and chorda, but no muscle or mesenchyme, and the vegetative embryo presenting the opposite structure.

When the section is made under the equator, the chances of normality necessarily increase for the upper, animal embryo, while the vegetative fragment always gastrulates imperfectly. Nevertheless, a neural plate or groove may be found in the ectoblast, and a chorda, muscles and mesenchyme between this layer and the partially naked entoblast. Such vegetative embryos are very similar to those formed of isolated macromeres (fig. 39, *c*). With a still lower section, the main embryo is perfect, the vegetative part yielding only entoblast, muscle, mesenchyme and a small piece of ectoblast (fig. 42, VI). The dimensions of the organs in this last case are given in Table V.

Table V

Embryo	Total size (%)	Ento-blast (%)	Epi-blast (%)	Neural organ (%)	Chorda (%)	Mesen-chyme (%)	Muscle (%)	Spaces (%)
An	75·5	24·5	40·5	8·5	6	9	12	0·5
Vg	24·5	85	3·5	—	—	3·5	7·5	0·5
Both	100	39·5	31	6·5	4·5	7·5	11	0·5
Controls	100	± 25·5	± 37·5	± 6	± 7·5	± 7	± 10·5	± 6

The *An* embryo is normal, except for its enlarged neural organ. *Vg* is incomplete, with a limited plate of ectoblast, without any neuralisation. Chorda is lacking, but mesoblast is present. The summation of the respective structures results in an exaggeration of entoblast and a reduction of epiblastic and chordal material.

The anatomy of these "latitudinal twins" immediately excludes the possibility that the unfertilized cytoplasm is homogeneous. Differences arise according to the original situation of the fragments, and they cannot be attributed to the fact that one embryo is haploid and the other diploid, for the bigger part of an equally divided egg develops into a normal tadpole, no matter if that part received the female pronucleus or not. On the other hand, it is not very likely that the development of the fragments corresponds to the colourable areas of neuroblast and chorda. Even with some discordance between the axis of maturation and of morphogenesis, a mosaic structure similar to that of fig. 40 would be expected to give rise to incomplete embryos, and complementary twins. Certain special dispositions also would hardly tally with the equatorial superposition of two determined neural and chordal crescents. Animal halves may form a large chorda with a small neural organ; from vegetative ones, a neural organ may arise, and no chorda. A case must also be mentioned where an animal half lacks a chorda, but has entoblastic cells swollen with big vacuoles. It is thus impossible to assume a mosaic development of the parts. In spite of the two remarkable cases of disjunction between neuro-chordoblastic and mesoblastic cells, the fate of the presumptive chordo-neural material at least is certainly changed in most merogonic embryos. Measurements of the organs confirm this opinion. In the rather frequent cases where one embryo lacks any neural organ, its brother has not the giant brain and medulla which would appear if it had simply received the whole neural area, but its neural organ is of normal size. The same is true, with one exception, for chorda. When two qualitatively complete embryos are formed, the ratio between epiblast and entoblast may be considerably altered, the first one being formed in excess in animal subhalves and being reduced in vegetative subhalves. The other organs are either more or less defective, or of normal proportion, but rarely exaggerated.

With *meridian sections*, the embryos are of equal size, but nevertheless present an instructive variety of structure. Contrary to what happens after separation of the two first blastomeres, the occurrence of hemiembryos is exceptional. I have obtained only two pairs of these, and a few isolated embryos from cases where the other fragment did not develop. One larva especially attracted attention by its paradoxical structure, the mesenchyme being located on the left, under the typically cicatricial epiblast, in an otherwise right hemiembryo; this again is scarcely compatible with a mosaic representation. It suggests that the formation of hemiembryos may be related to the extent and quality of the new-formed film. This is confirmed by the occurrence of pairs, one member of which is a normal tadpole, while the other is a hemiembryo (fig. 43, I). In this case, the normal embryo is also curiously distorted and somewhat asymmetrical (same drawing, right). The dimensions of the organs are indicated in Table VI. Except for mesenchyme, which is reduced to about half of its normal amount, the quantity of each organ is very close to the normal figure for a tadpole of the same stage.

Table VI

Embryo	Total size (%)	Ento-blast (%)	Epi-blast (%)	Neural organ (%)	Chorda (%)	Mesen-chyme (%)	Muscle (%)	Spaces (%)
a	31	30	29	10	5·5	2	9	14·5
b	69	22·5	32	5·5	5·5	5	8	21·5
Both	100	25·5	30·5	7·5	5·5	3·5	8·5	19
Controls	100	± 25·5	± 37·5	± 6	± 7·5	± 7	± 10·5	± 6

Other results of meridian sections fall under two main heads. In certain pairs, the two embryos are different one from the other (fig. 43, II). One is fairly normal, the other has a very stumpy tail, sometimes with the fusion of the left and right muscles. Chorda and neural organ may be reduced in the abnormal embryo, the latter even being absent. An examination of the dimensions attained by the various organs (Table VII) shows that in this case the disturbance exclusively bears on epiblast and the neural organ; other variations are within the limit of error of these measurements.

In other pairs, the resemblance of the twins is extreme, as much for the form (fig. 43, III) as for the size of the organs (Table

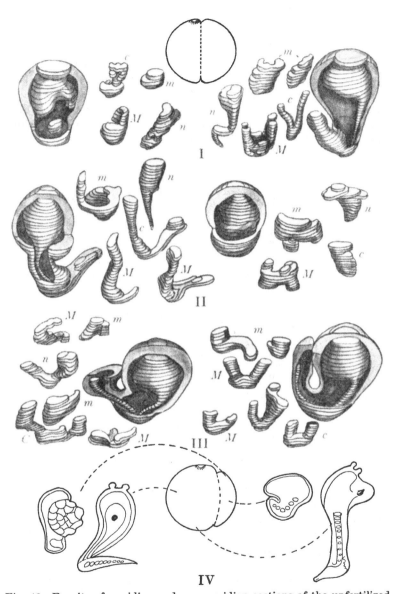

Fig. 43. Results of meridian and parameridian sections of the unfertilized egg of *Ascidiella scabra*. I, II, III, Meridian sections. In each case, the reconstruction shows, for both twin embryos, the entoblast placed in the epiblastic envelope, and the other organs separately represented, in a logical disposition. I, Hemiembryo (left) and distorted complete embryo (right). II, Normal embryo (left) and anouran embryo (right). III, Two mirrored embryos. IV, Two experiments of parameridian section, after free-hand sketches of the living embryos. Abbreviations as in fig. 39.

VIII). But the embryos are not really normal. Both are strongly
bent backwards, and show a torsion of the tail in opposite directions; they are sometimes, as in the case illustrated, completely
mirrored. The relative size of the organs is perfectly normal,
except for a distinct excess of epiblast; this is of course unavoidable, the surface to be covered being larger.

Table VII

Embryo	Total size (%)	Ento-blast (%)	Epi-blast (%)	Neural organ (%)	Chorda (%)	Mesen-chyme (%)	Muscle (%)	Spaces (%)
a	48	21·5	46	5	4	10	9·5	4
b	52	22	45	5·5	5·5	6	8·5	7·5
Both	100	21·5	45·5	5	5	8	9	6
Controls	100	24·5	35·5	7·5	6	8	10·5	8

Table VIII

Embryo	Total size (%)	Ento-blast (%)	Epi-blast (%)	Neural organ (%)	Chorda (%)	Mesen-chyme (%)	Muscle (%)	Spaces (%)
a	50·5	20·5	40	7	8	8·5	8·5	7·5
b	49·5	20·5	41	6	6	7·5	7·5	11·5
Both	100	20·5	40·5	6·5	7	8	8	10
Controls	100	25·5	37·5	6	7·5	7	10·5	6

The experiment has been repeated a sufficient number of times
to assert that a pair of perfectly normal embryos cannot be
obtained. Regulation has a limit, as much with meridian as with
latitudinal sections. These results have again a negative and
positive meaning. They are certainly incompatible with a mosaic
organization of the cytoplasm. If the factors responsible for
chorda-differentiation were only present in its small presumptive
crescent, about half of the operations would result in one of the
twins being deprived, or nearly so, of chorda. If the factors for
the neural organ were located in its long-horned presumptive
crescent, all embryos would be provided with it, but in half of
the cases both twins would possess brain and medulla, in the
other half one embryo would have a large, mostly cerebral,
nervous organ, its brother only a thinner, medullar tube. Neither
of these anticipations appears to be correct. Chorda never fails,

and its reduction in one individual is exceptional. The neural organ, on the contrary, is often either absent, or deficient in one member of the pair, and, in such cases, its volume is not increased in the twin-embryo. With regard to the positive information of these meridian sections, the most important data are that the hemiembryos are exceptional and in some cases of a paradoxical structure; that two perfect tadpoles are not obtained; that certain pairs are formed of similar members, others of different ones. These results can be best understood by the assumption, in the unfertilized egg, of some previous bilateral structure. The deficiencies are to be attributed to the orientation of the section, the plane of separation having been respectively frontal (dissimilar embryos) or sagittal (similar, sometimes mirrored embryos), or oblique (intermediate cases).

The third series of experiments, the *section along a parameridian plane*, may be accounted for very briefly. A sphere separated from the equatorial region of the egg, and amounting to one-fourth or one-fifth of its volume, develops into a slightly elongated mass, where all organs can be cytologically recognized. The constant presence of chorda confirms the previously mentioned hypothesis that the factors responsible for its formation cannot have the narrow localization indicated by the vital staining observations. The structure is, however, quite atypical, approaching that of a hemiembryo. The large fragment of the egg gives rise to a tadpole with a more or less normal body but a variable tail (fig. 43, IV). One case could be ascertained, thanks to a definite alteration of the epiblast in its antero-dorsal region, to have been cut in the dorso-median region. The anterior part of the chorda is embodied in the thickened epiblast, while the neural organ is divided into two cerebral vesicles, a left and a right. This disposition again cannot be imagined as a transformation of a preformed mosaic of specialized plasms. On the other hand, it cannot be the result of a homogeneous mass of cytoplasm. Between these two extreme representations, an intermediate state of organization must be found, sufficiently supple to account for the data of observation and experience.

The three groups of experiments made on the Ascidian egg before its fertilization thus agree in rendering untenable a true mosaic conception, and provide sufficient evidence for us to deduce the real **internal organization of the cytoplasm** in the unfertilized Chordate egg.

The classical terminology of *prospective significance and prospective potency*, introduced by Driesch, is much more subjective than it appears to be. It supposes, or suggests, that a part of the egg acquires, when isolated, properties which it did not possess before the experiment. I feel it advisable to avoid such a conception. The potencies exhibited by an isolated part of a germ are only the expression of the organization inherent to that part. It has, of course, undergone a change of form and of relative surface. But this is the inevitable consequence of isolation. Provided no mechanical injury has happened, the morphogenetic activity is the true expression of the internal organization. So long as we are not compelled by evidence, we must refuse to invoke latent forces. Experiments teach us that development remains normal while such and such elements are present, and becomes abnormal in certain other conditions. The internal organization of the Ascidian egg is the necessary and sufficient representation to account for these facts. Let us try to define, in the case of the Ascidian egg, what and where are the factors responsible for the formation of the primary organs.

The material for mesoblast is, as already stated, distinct from the beginning as a part of the cortical plasma. In its animal part, this superficial film is free of the mitochondrial granules which bear the yellow pigment. This is also the region which is unable to form muscular cells and typical mesenchyme. The potency to build up these elements is bound to the yellow-pigmented plasm, with a reduced activity for the most vegetative parts of it. Attention must also be paid to the elongated form of the animal region free of pigment,[1] as a sign of bilateral symmetrical organization. No definite evidence could be found for distinct plasms forming respectively mesenchyme and muscle. The segregation probably happens after fertilization according to reactions soon to be discussed.

The neural and chordal organs are normally formed by the cytoplasm of their two contiguous but unequal presumptive

---

[1] Which also exists, even more conspicuously, in *Styela*; no elongation has been noticed by Conklin.

crescents (fig. 40). This material may, however, find another utilization, as is shown by various cases of important reduction in the total amount, for both twins, either of chorda or of neural organ; or sometimes of both. Again, other material is able to take the place of the normal substrate of those organs. We could, of course, imagine an extension of each presumptive crescent into a broader potential field, with a centrifugal decrement of the specific factors. But it would not explain why chorda and neural organ are not always obtained together, in correlated proportions; why the more animal material at one time forms chorda, at another neuroblast, and both in a third case; why the same anomalies may be encountered in a vegetative part. In a word, when we attempt to localize chorda and neural areas separately, in the unfertilized egg, we stumble on paradoxical situations. If a distinct location appears impossible, we have to assume a unique condition for the two dorsal organs, and suppose that secondary factors decide that this specialized material will become chorda, neural organ, or both.

Normal development gives us a hint concerning that secondary influence: the grey crescent of *Styela* with its relatively late appearance, its homogeneous aspect, its ulterior division, by the latitudinal furrow, into the micromeres and the macromeres. Its superior part lies in the ectoblastic territory and, from the beginning of gastrulation, it shows, when vitally stained, a special, somewhat granular aspect.[1] The lower part is included in the invaginating material and is the appanage of the chordal territory. In short, given the grey crescent, a correlation similar to that demonstrated for Amphibians appears possible. Gastrulation, which can hardly be imagined to depend on less than two distinct factors, may be assumed to result, in Ascidians also, from interaction between a polar gradient and a peripheral field, the most active part of which lies in the grey crescent. The amount of the x-substances produced, which constitutes a cortical field of morphogenetic potentials, produces epiboly or invagination, according to whether it is above or below a definite threshold. The initial general flattening of the blastula, the greater activity of the dorsal grey region, particularly the intense power of extension exhibited by the chordal cells, thereby receive their

---

[1] Oral communication of G. Vandebroek.

explanation. Again, the quality of chorda will appertain to the invaginating material initially charged to a certain extent with the grey plasm, and, similarly, the neural quality to the superficial cells which have a similar content. It may also be assumed that the mesoblast results from the existence, in certain of the invaginating cells, of a sufficient amount of mesoplasm, a subsidiary threshold allowing the distinction between muscle and mesenchyme.

Quantitative variations in the interaction of a cortical field and of a fundamental, animal-vegetative gradient, extended to the whole body of the cell, are thus thought to account for two kinds of results; first, the general property of epiboly and invagination, second, the determination of the four primary organs: neural organ and chorda, muscles and mesenchyme. The other two, epiblast and entoblast, are simply the parts of the respectively epibolic or invaginating materials, which are the least charged with the "grey" principle, and also, for the entoblast, deprived of mesoplasm. Although the evidence is less complete than in Vertebrates, gradient and field may be thought of in the same terms. The former pervades the whole germ and is only a new expression of the well-known primary polarization. To call it a "yolk gradient" would not be entirely justified in the case of *Ascidiella*, where the distribution of the yolk platelets is extremely uniform before cleavage. But there is, as indicated above, other cytological evidence of a polarization which is more conspicuous in the *Amphioxus* egg. The existence of the peripheral field is of course admitted by analogy with the facts offered by the Amphibian egg (p. 83). In any case, some prepared structure of the Ascidian egg is necessary to account for the migration of the yellow plasm on the ventral side, and for the appearance of the dorsal grey crescent. The peripheral, dorsoventral field fulfils this condition. It also affords a satisfactory explanation of the rather complex results obtained by double merogony. The sections of the egg may evidently affect the gradient and the field in various degrees, according to the direction of the cut. But if the field is inherent to the surface film, a consequence of any section will be that this morphogenetic factor will be reduced, by the unavoidable stretching of the film, in the region of the section. And this is a fact of general observation, as much in latitudinal as in meridian and in para-meridian sections.

If the reactions governing gastrulation and organogenesis exert different effects according to definite thresholds, many combinations are able to appear in merogony experiments. The amount of dorso-ventral "plasm" may be sufficient for gastrulation, but not for chordal or neural organogenesis, or only for one of these. As a matter of fact, all degrees of differentiation are encountered. The neural segregation may be either null, or slightly indicated, or more complete as an organ but not cytologically. The chorda may be either scarcely recognizable, or well isolated as an organ, but with imperfect vacuoles, or perfectly formed. Small animal vesicles may exhibit incompletely formed mesenchyme, etc. The variety of the observed results becomes still more intelligible when one remembers the variable discrepancy between the polar spot and the animal pole, the latter being determinated by the exact position of the focus of the dorso-ventral field relatively to the polar gradient.

The behaviour of the four micromeres and of the four macro-meres, when separated, combined or translocated (p. 104), now finds its explanation. In the first operation, the gradient and the field are both affected in their equatorial region. We do not know if the gradient is modified or not, but the field is certainly poor on the previous interface between the micro- and the macro-meres. Its new focus corresponds with the grey material of the presumptive neural area. About this level, the supposed inter-action attains a "morphogenetic potential" surpassing the threshold of invagination and gastrulation takes place with a limited entoblast. In the isolated macromeres, the situation is the opposite. As to the organs formed from these atypical gastrulae, it is to be noticed that chorda and muscle only exceptionally appear out of a simple or double group of micromeres; the interaction has already acquired a sort of neural orientation. Conversely, macromeres do not yield a neuralized ectoblast in all cases, but its occurrence (fig. 39, c) indicates that the determination is not yet settled. The strongest impression of determination emerges from the 180° rotation of micromeres (fig. 39, d, e), precisely because the internal gradient is not appreciably changed. The cortical field alone is divided horizontally and its focus shared between the dorsal and the ventral region. Gastrulation begins normally, but it is probably achieved in a very peculiar way, which requires further study, and a tadpole is obtained with an apparently reversed neural organ. The importance of a subequatorial region of the fertilized egg for a normal cleavage and a good conformation of the embryo may also be related to the position of the morphogenetic dorsal focus.

New researches, indeed, are required on many points of Prochordate development. In our present state of knowledge, the theory just arrived at accounts for the normal and experimental features of development, without any call on special processes, latent properties or anything of that kind. The frequent occurrence of regulation in operations performed on the unfertilized egg, its failure in definite cases, the apparent determination of ventral and dorsal blastomeres at the 4-cell stage, the relative plasticity of the 8-cell stage, all these well-analysed facts now receive their interpretation.

Between Ascidians and *Amphioxus*, the main difference, from the present viewpoint, is in the result of the isolation of the first two blastomeres. While the surface of previous contact remains morphogenetically inactive in Ascidians, it shows a nearly normal developmental ability in the other species. Is this due to the extension of the dorso-ventral field along the edges of the furrow, or to the secondary diffusion of the assumed substances of the morphogenetic field? At present the question can only be asked.

Between Prochordates and Vertebrates, two main differences are perceived. First, the former phylum possesses a special plasmatic differentiation, the mesoplasm, while in the latter, the segregation between chorda and mesoblast depends on the primary ratio between the yolk gradient and the dorso-ventral field. Secondly, the segregation of ectoblast into neuroblast and epiblast seems to be due, in Prochordates, to intrinsic conditions; in Vertebrates, on the other hand, to an induction. In the latter case, the areas where the products of the $C. V$ reaction are under the invagination threshold lack *ipso facto* the condition for neuralization. Thanks to the invagination of richer material, a part of these products, the organizin, is able to pass into the overlying ectoblast, and the critical threshold is there attained and surpassed. The condition directly realized in Prochordates is thus attained by a roundabout way. For secondary segregations, however, striking similarities are observed: the inductive action of the prechordal material on the adhesive papillae in Ascidians, on the stomodoeum, in Vertebrates, then on the sense organs produced in both cases by the brain, offers the most remarkable parallelism.

A last remark before leaving the Chordates. The present account has not taken into consideration the **problem of the visceral asymmetry** secondarily appearing in all representatives of that large phylum. This feature has, however, been demonstrated to be modifiable by experimental means as much in the *Amphioxus* (p. 103) as in the newt.[1] It is certainly not acquired under environmental conditions, and must be represented in some way in the germ at most early stages. The suggestion that there is some corresponding asymmetry in the dorso-ventral field, or in the yolk gradient, and consequently in the field of morphogenetic potentials, is really too easy and lacks any objective basis. It is more advisable, in my opinion, to state that this process is not concerned with the functions that we are particularly studying in this book. The establishment of the *situs viscerum* is not the consequence of morphogenetic movements. It proceeds essentially from a differential growth which follows a definite mathematical law. It is related to the large problem of *growth gradients*, which play such a part in the definitive establishment of the proportions between the organs, and in their detailed moulding. Their importance has been rightly emphasized[2] in the processes of development and regeneration. But the nature of their functional bases and their linkage with the early morphogenetic processes are not yet elucidated.

[1] Spemann and Falkenberg, 1919.
[2] Huxley and de Beer, 1934; Huxley, 1935.

CHAPTER IX

Field and gradient in the sea-urchin egg

Comparison has always been an important method of progress
for biological sciences. It certainly helps a great deal in causal
Embryology. In spite of the obvious differences in the organization of the zoological types, the ultimate causes of morphogenesis
are probably of a similar nature among far-related species, and
their discovery calls for a constant survey of all cases where early
development has been submitted to methodical study. The sea-
urchin egg occupies, in this programme, a privileged position,
for its analysis has been carried very far, with a still larger use
of the quantitative and biological point of view. Under the
competent guidance of J. Runnström, a group of young cyto-
physiologists has brilliantly solved important problems; the
clever micrurgical experiments of Hörstadius have been decisive
factors for these advances.

The **polar organization of the egg** is, in the sea-urchin, marked
less by the position of the nucleus in the oocyte than by cytoplasmic features. In one case at least, a girdle of pigment
surrounds the equator of the virgin egg. After fertilization, the
third cleavage forms a latitudinal furrow more or less superior
to the equator (fig. 44, c). The subsequent mitotic cycle causes a
group of micromeres to appear at the vegetative pole. They give
rise to small cells endowed with a special activity. They are the
first to invade the blastocoele (fig. 44, h), where they form two
masses of primary mesenchyme (i), which produce the larval
skeleton (i, j). Simultaneously, the true gastrulation takes place
in such a way that the top of the archenteron curves slightly
between the two skeleton-forming masses (k). This inflexion
indicates the future dorsal side. Thus, from the gastrula stage,
a dorso-ventral axis, more or less perpendicular to the polar line,
is conspicuous; a plane of bilateral symmetry is thereby trace-
able. From that stage, the extension of the various regions,
especially in the enveloping ectoblast, exhibits local differences,

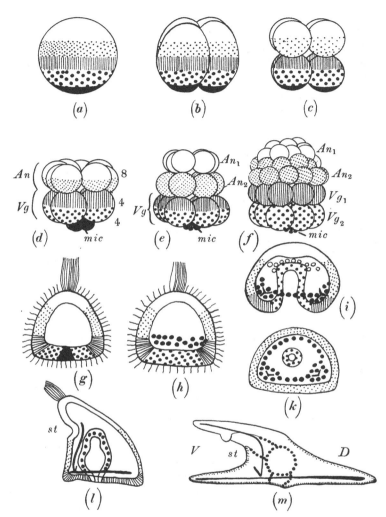

Fig. 44. Schematic representation of the larval development in the sea-urchin, especially to show the primitive position of the materials destined to each region. In black: the micromeres; coarse dotting: entoblast, secondary mesenchyme and coelomic material, localized in Vg_2 during the 64-cell stage (f); vertical hatching, light stippling and white: ectoblastic areas respectively located in Vg_1, An_2 and An_1 during the 64-cell stage (f); D, dorsal; V, ventral; st, stomodoeum. Redrawn after Hörstadius, 1935; modified.

which also affect the cytological structure of the epithelium. After a prismatic stage, the polar axis becomes progressively bent (*l*): the material of the original animal pole moves ventrally, while the cul-de-sac of the archenteron inclines in the same direction, approaching the small stomodoeal depression which has hollowed out meanwhile (*l*). Secondary mesenchyme and the coelomic anlage are also formed by cells proceeding from the archenteric wall. It is not necessary to elaborate here on the formation of the larval arms and of the peristomodoeal ciliation, on the shape of the skeleton pieces, and on the division of the digestive tract in oesophagus, stomach and intestine, in a word on the anatomy and histology of the normal pluteus (*m*). It is enough to recognize, from the stages of larval development, that the analysis of the egg organization must successively consider the animal-vegetative axis, the bilateral symmetry, and the dorso-ventral differentiation. The later appearing asymmetry and the transformations of metamorphosis may be left out of the scope of this study.

From the pioneer work of Herbst, it is well known that the larval morphogenesis is, in the sea-urchin, extremely **sensitive to the chemical composition** of the environment. Exo-gastrulation provoked by lithium is the most typical example, and, in its most pronounced form, it certainly implies a real hypertrophy of entoblast at the cost of other materials; this means indeed an exaggeration of the structures characteristic of the vegetative part. In quite different experiments, when separating the sea-urchin blastomeres, Driesch, on the contrary, had noted the formation of hollow vesicles provided with long stiff cilia; this structure is an exaggeration of the animal ciliary tuft of the advanced blastula (fig. 44, *g*). These two quite opposite anomalies allow us to make the distinction between processes of *vegetativization* and of *animalization*. Numerous experiments, both chemical and micrurgical, have demonstrated that each of these extreme deviations is the end-term of a series. Animalization, when only slightly indicated, simply consists in a reduction of the enteron; when it is more marked, the larval arms are diminished or suppressed; then the stomodoeum, which proceeds from material located not far from the animal pole, disappears; then again, the ciliary band that surrounds the oral field of the young pluteus is reduced; finally, we arrive at the suppression of all organogenesis, although some epithelial differentiation still indicates the dorso-ventral orientation; the extreme term is the expansion of the cilia—normally limited to the apical tuft—to still more vegetative parts and the increasing elongation of these abnormal plasmatic processes (fig. 45). Vegetativization, on the contrary,

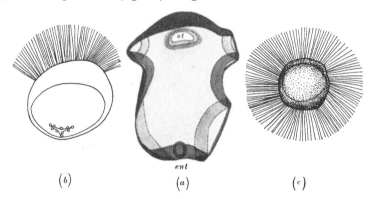

(b) *ent* (a) (c)

Fig. 45. Three stages of animalization in eggs submitted, before fertilization, to NaSCN in Ca-free sea-water. (*a*) Larva with extreme reduction of the entoblast (*ent*), persistence of the stomodoeum (*st*), and considerable ciliary band; the dorso-ventral polarization is quite distinct; ventral view. (*b*) Suppression of any organ except a tiny mass of mesenchyme with a spicule; extension of the ciliary tuft. (*c*) Extreme case of an hollow vesicle with uniform long and stiff cilia. Redrawn from Lindahl, 1936.

is first indicated by hypertrophy of the enteron, which is then less deeply invaginated; when further advanced, it suppresses the stomodoeum, which depends on both animal and vegetative activities, and makes the arms shorter; a step further, and the entoblast evaginates, appended to an ectoblastic vesicle containing some calcarous spicules (fig. 46); in extreme cases, the entoblast, turned inside out like a glove, absorbs the primary mesenchyme and a part of the ectoblast, of which a very small portion only remains. There are certainly, in this very rich series, numerous other intermediate types. It is sufficient, for our purpose, to be acquainted with the general appearance of these induced anomalies. What is their causality?

132   FIELD AND GRADIENT

They are, firstly, to be induced, as mentioned, by a change in the environment. Vegetativization is most regularly obtained by the cation Li+, this being added to normally fertilized eggs, or to virgin eggs in sea-water without calcium.[1] It is not a specific action, and many other conditions[2] give the same result: butyrate of Na, NaCl, $HgCl_2$, $HClO_3$, $CuSO_4$, KCN, various other salts and salt mixtures, the lack of magnesium, high and low temperatures, neutral red, methylene-blue, auxin, glycogen, tobacco-smoke,

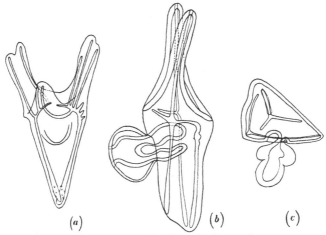

(a) (b) (c)

Fig. 46. Three types of vegetativization. (a) Larva from an egg treated, before fertilization, by lithium in Ca-free sea-water; incomplete invagination of the entoblast. (b) Egg submitted to lithium in otherwise normal sea-water during the first 24 hours after fertilization; partial exogastrulation. (c) Same treatment; the whole entoblast is evaginated, while ectoblast and skeleton are relatively unaltered. Redrawn from Lindahl, 1936.

X-rays. The efficiency of lithium is nevertheless remarkable, and it has been possible to study its effects on the metabolism. By numerous measurements and skilful comparisons, the conclusion has been arrived at that the metabolism, in the animal part of the egg, is principally the consumption of carbohydrates, and that their utilization is more or less inhibited by Li+.[3] Other results also indicate an excitation, by the same agent, of the vegetative metabolism,[4] which seems to affect mainly proteid

[1] Lindahl, 1936, p. 194.   [2] Child, 1937, p. 459.   [3] Lindahl, 1936.
[4] Lindahl and Öhmann, mentioned in Lindahl and Stordal, 1937.

substances. It has been shown that the necessity of SO_4^- anions in sea-water is connected with the neutralization of certain toxic, probably aromatic, products of protein disintegration, which only takes place in the vegetative part of a normal egg or in its equivalent.[1] Conditions provoking animalization seem to be less numerous. It can be induced by treating the virgin egg with Ca-free sea-water, to which has been added either NaSCN, or NaI, or pyocyanin. After washing the eggs with normal sea-water, fertilization takes place normally and animalization appears in variable degrees. By its depressive action on the vegetative activity, the lack of sulphate is equally able to cause a slight degree of animalization. It should also be mentioned that these artificial conditions generally affect the respiratory metabolism, the addition of Li^+ and the lack of sulphate having both a distinct and depressing action on the oxidations. Probably there is here a hint for a future elucidation of the relation between respiratory metabolism and morphogenesis, but, as in Amphibians, this important possibility can at present only be indicated.[2]

The idea of combining the two methods immediately suggests itself. Eggs previously submitted to the animalizing treatment may be, after fertilization, exposed to Li^+. The effect is sometimes simply a corrective one, but it may also provoke the formation of an extraordinary structure. The vegetative pole is typically animalized, but an equatorial ring acquires the ability of forming primary mesenchyme and later becomes stretched into a kind of enteric tube connecting two ectoblastic vesicles which contain calcareous spicules and eventually acquire the appearance of defective plutei (fig. 47). The interpretation of this result is not easy, but it is, in any case, a testimony of the immense morphogenetic plasticity of the sea-urchin egg and of the action of physico-chemical conditions on its organization.

[1] Lindahl and Stordal, 1937.
[2] Child (1937) has shown that various sea-urchin eggs, when coloured with methylene blue or janus green and then deprived of O_2, present, up to the blastula stage, a gradual discoloration from the animal to the vegetative pole. When primary mesenchyme begins to be formed, this basipetal gradient is complicated by another system starting from the migrating cells and becoming more or less acropetal. These facts are probably correlated with a respiratory decrement along the axis, a conception supported, to a certain point, by the data concerning the animal and vegetative metabolism.

Although more tedious and delicate, operative methods have led, in this case also, to most remarkable achievements. They are specially based on the possibility of **isolating the discs of blastomeres,** which are found superposed, during cleavage, from the

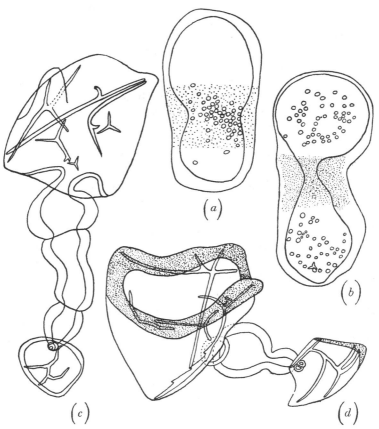

Fig. 47. Combined effect of chemical animalizing and vegetativizing treatment. Eggs treated with NaSCN in Ca-free sea-water before fertilization, and then with Li+ after fertilization. (*a*) Aspect at age of 1 day. (*b*) Aspect at 1½ days. (*c*), (*d*) Abnormal larvae aged 3 days. From Lindahl, 1936.

vegetative to the animal pole. For technical reasons, these operations cannot be conducted at one definite stage; the possibilities offered by the successive stages must be conveniently utilized. At the 8-cell stage, the latitudinal furrow, slightly super-equatorial, separates the so-called animal and vegetative territories (fig. 44, d), and it is possible to isolate these one from another without any damage, if the egg has been placed in Ca-free sea-water. During the following cycle, the spindles are orientated and situated in such a way that the animal part forms a rosette of 8 mesomeres superposed on the vegetative material, which is divided into 4 macromeres and 4 micromeres (fig. 44, d). This stage, which Hörstadius represents by the formula $8+4+4$, is especially favourable for the isolation of the superior group of 8 mesomeres, and of the inferior 4 micromeres. The mesomeres then divide latitudinally into two discs of 8 cells, called An_1 and An_2; the macromeres divide, their spindles being horizontal, into a group of 8 cells, still superposed on the micromeres. This 32-cell stage (formula: An_1+An_2+4+4) thus allows, in Ca-free sea-water, the separation of the discs An_1 and An_2 (fig. 44, e). Another mitotic cycle is necessary before the daughter cells of the macromeres divide, this time latitudinally, into two octets Vg_1 and Vg_2. Thus, operating on a 64-cell stage (formula: $An_1+An_2+Vg_1+Vg_2+mic$) it will be possible with the technique of Hörstadius[1] to isolate Vg_1 and Vg_2 (fig. 48, f). These relations are summed up in the following scheme:

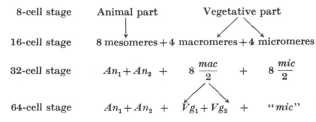

This state of affairs thus provides the means of isolating the discs constituting the egg and of freely combining them, or some of them, in numerous ways. It seems that all significant combinations have been actually performed. To grasp their importance, it is necessary to be exactly acquainted with the normal fate of the various parts of the egg, and to localize exactly the limit of invagination. While it was soon recognized that the primary,

[1] 1935, 1936.

skeleton-forming mesenchyme proceeds from the micromeres, it has been long and difficult to trace the ento-ectoblastic boundary. It is now established that it lies between Vg_1 and Vg_2, as is to be deduced from the two experiments represented by fig. 48. The same observations, repeated on blastomeres of other discs, determine the presumptive fate of each one of the superposed slices (fig. 44).

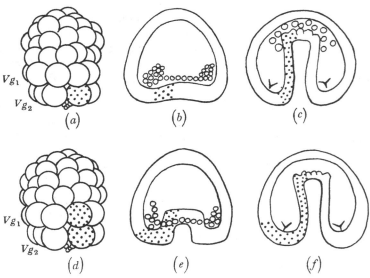

Fig. 48. Vital staining of vegetative blastomeres at the 64-cell stage. (a), (b), (c) One Vg_2 blastomere has been coloured and its material is exclusively entoblastic. (d), (e), (f) Two superposed blastomeres, belonging respectively to Vg_1 and Vg_2 have been coloured, and their material is partly ectoblastic, partly entoblastic. Redrawn from Hörstadius, 1936.

The evolution of the same materials is absolutely different when they are isolated during cleavage. The An_1 disc forms an hyperciliated blastula, a typical manifestation of animalization. An_2 tends to a reduction of ciliation, forms a ciliary band, sometimes a stomodoeum. Embryos developed from Vg_1 are not uniform; sometimes, they are mere ciliated vesicles, such as are also yielded by An_2; in other cases, they possess a short archenteron, sometimes a stomodoeum, and certain larvae even present a pluteus-like shape, but with an imperfect skeleton. With Vg_2, an accentuation of the vegetative activity can be clearly observed; no ciliary tuft is ever formed; a broad archenteron invaginates, and becomes constricted into two or three parts; the mouth cavity is always absent; the ectoblast shows no differentiation other than a ciliated field; the skeleton is rudimentary, with one or sometimes two spicules. Finally, the micromeres themselves seem to be deprived of any intrinsic morphogenetic power; they yield nothing but a mass of round cells.

This inertness of the micromere material is however a fallacious impression, as will soon appear from the study of **compounds made of various discs.** Firstly, what are the potencies of those natural compounds represented respectively by the animal and the vegetative part? In most cases, the former does not give more than a slightly irregular vesicle, with a ciliated band and a stomodoeum (fig. 49). With certain eggs, however, and using

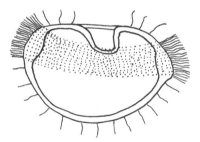

Fig. 49. Development of an isolated animal half. Redrawn from Hör-stadius, 1935.

an appropriate technique, nearly normal plutei may be obtained.[1] The vegetative half often shows an almost complete organo-genesis, but the embryo is generally abnormal, although perfect in some cases. As will appear from this survey, the observed variations may be attributed to a certain shifting of the latitudinal furrow towards one or the other pole.

The important result of combinations between the various discs isolated from the egg can be summed up in a few words. Given a certain mass of animal material, the development will be the more normal the greater the amount of added vegetative

[1] Cf. v. Ubisch, 1936.

substance. An_1 requires a stronger compensation than An_2; on the other hand, Vg_1 has a less vegetative effect than Vg_2, which itself is surpassed by the micromeres. When the quantity of vegetative material becomes too strong, all stages of vegetativization are to be observed. That a real quantitative relation is at work here is shown by the comparable effectiveness of either 1, or 2, or 3, or 4 micromeres added to one of the principal tiers of the egg. For An_1 and An_2, the equilibrium necessary to the formation of a pluteus is reached with Vg_1 and Vg_2, the addition of more vegetative material soon resulting in exogastrulation.

At first sight, the effects of isolating the discs gives the impression of a fundamental change in the fate of the egg materials. This is really, however, not so frequent as it appears. When the animal part of the egg forms a hyperciliated blastula, the ectoblast is not modified as a layer, but merely in its differentiation, because it is liberated from the influence normally exerted upon it by the more vegetative regions. When the vegetative half develops by itself into an abnormal larva, there is, nevertheless, no displacement of the limit between ectoblast and entoblast, and the number of cells migrating into the primary mesenchyme is also not increased; the archenteron utilizes the whole material of the presumptive entoblast, and consequently becomes relatively too large.[1] Some cases, however, involve a displacement of the limit between the external and internal layers. When the excess of ectoblast is considerable, as in compounds $8+2+2$, $8+1+1$, $8+1+0$, i.e. the mesomeres added with a half, a fourth or one-eighth of the vegetative part, the limit of invagination becomes displaced upwards, and the archenteron is partially built of presumptive ectoblast. Conversely, Vg_2, when cultivated alone, forms a piece of ectoblast, a development which also implies an essential change of fate for the animal part of this disc. We must, however, beware of believing that such processes always involve an effort towards normality. In cases where the vegetative material is less than in those mentioned above, e.g. in $8+\frac{1}{2}+0$ or $8+0+1$, this material, instead of increasing, is simply assimilated into the ectoblast. Generally speaking, the displacement of the limit between ecto- and entoblast may be said to be exceptional; it has a real tendency to maintain its normal level.

[1] Hörstadius, 1936 c, p. 47.

When the equilibrium is too much disturbed, a kind of contagion takes place in the one or the other direction, but the process is deprived of any teleological signification.

This being admitted, it becomes easy to define the nature of the changes observed in the isolated discs, or in artificially combined systems. It is clear that if $An_1 + An_2 + Vg_2$ builds up a larva provided with a skeleton, this means that the most vegetative

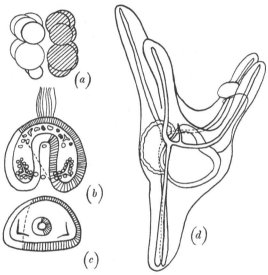

Fig. 50. Experiment of joining a meridian and an animal half. (a) Scheme of the experiment, the animal half right. (b) The gastrula; the part proceeding from the animal fragment is shaded. (c) Optical section of the same larva, viewed from the animal pole. (d) The resulting normal pluteus. From Hörstadius, 1935; redrawn.

part of Vg_2 has taken over the function of the absent micromeres; and the same remark holds good when $An_2 + Vg_1 + Vg_2$ forms a normal pluteus. Moreover, in the latter case, the polar ciliary tuft of the gastrula stage must be produced by cells which normally lie in normal ectoblast; they thus replace the most animal part of An_1. Again, let us consider, in order to realize the importance attained by these changes, the fine experiment where the animal half of one egg is placed against the meridian half of another (fig. 50). A previous vital coloration shows that the eight added mesomeres contribute to the formation of the archenteron as well as of the ectoblast and thereby permit the organization of a perfectly normal pluteus.

The rule which appears from these experiments is that, in any sufficiently equilibrated complex, the vegetative and the animal gradients—this term being used in its most general sense—concentrate again in the direction of their respective poles. We will try later to give to this conclusion a more suitable and precise expression. As such, it makes intelligible the fact that each of the blastomeres, in the 2-cell and even in the 4-cell stage, is able to build up a normal pluteus; it is to be predicted that the same will happen should the morula be skilfully divided into meridian sections. In point of fact, a 32-cell stage can, after elimination of the too-sticky micromeres, be divided into eight meridian portions, each of which is capable of developing into a more or less typical pluteus[1] (fig. 51). A still more direct verification of the conclusion arrived at is the obtaining of two plutei from an egg after division by simple latitudinal planes. Two kinds of operations lead to this result: after gradual division of an egg in its five discs, the group $An_1 + 0 + 4$ and $An_2 + 4 + 0$, or $8 + 4 + 0$ and $0 + 4 + 0$, may be aggregated. In both eventualities, a skilful operator gets in this way two plutei, slightly different in size, but remarkably normal[2] (fig. 52).

The matter is thus firmly established, and we have only to ask ourselves what is the variation, during development, of the potencies so clearly demonstrated for the early period of segmentation. For more advanced stages, information is easily provided by cutting at various latitudinal levels. The tendency to attain their normal fate appears to become still firmer as the age increases. In the animal half, especially the apical tuft, the ciliary band and the stomodoeum successively show their capacity of autonomous development. Another way of testing the stabilization of the potencies consists in the grafting of micromeres. It has been shown, and we shall have to consider this important fact again later, that these small cells are able to induce the formation of an archenteron in an ectoblastic territory. Now, micromeres may be grafted into animal halves isolated at con-

[1] Hörstadius and Wolsky, 1936, p. 87.
[2] Hörstadius, 1936 c, p. 61.

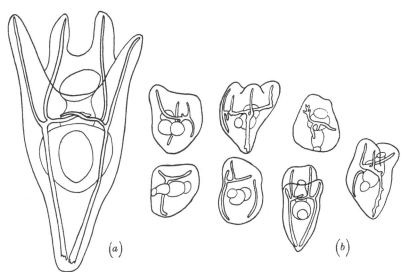

Fig. 51. Extreme case of regulation by division of the sea-urchin morula along meridian planes. The micromeres of a 32-cell stage have been eliminated; then, the germ has been divided in eight portions, each formed of a half macromere with the two overlying cells of An_1 and An_2. (a) A control pluteus. (b) The seven dwarf plutei obtained (the eighth was accidentally damaged). Redrawn from Hörstadius and Wolsky, 1936.

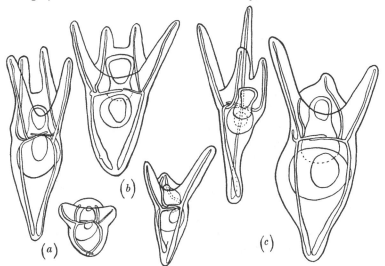

Fig. 52. Three pairs of plutei obtained by latitudinal separation of the discs of an egg and suitable re-combination of these in two groups. (a) and (b) $An_1 + 0 + 4$ and $An_2 + 4 + 0$. (c) $8 + 0 + 4$ and $0 + 4 + 0$. Redrawn from Hörstadius, 1936.

tinually increasing ages, and it is found that their inductive effect gradually vanishes. The length of time elapsed between the isolation and the graft has also its importance; a delay between the two operations allows some change to take place in the isolated fragment, with the result that, for equal ages, the susceptibility to the micromeres is distinctly reduced.

The stabilization of the potencies along the primary axis of the egg having been thus followed up to the late blastula stage, the inquiry has to be completed by examination of the **unfertilized egg**. The data concerning this important point were, until very recently, quite conflicting. In *Lytechinus* (*Toxopneustes*) it has been asserted[1] that merogony provided a normal larva in every case, whatever the orientation of the cut re-

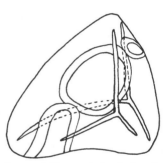

Fig. 53. A prismatic larva of *Paracentrotus l.* developed from a vegetative half isolated before fertilization. Redrawn from Hörstadius, 1928.

latively to the axis indicated by the micropyle. In *Paracentrotus*, sections near the nuclear area of the cytoplasm were supposed to demonstrate the rôle of the juxtanuclear material in the formation of micromeres.[2] But, on the same species, Hörstadius had established that eggs reliably orientated by means of a rosy pigment girdle show a distinct difference between the animal and the vegetative half. The animal half develops into a hyperciliated blastula, while the vegetative part forms a satisfactory pluteus (fig. 53). The same experiment of equatorial

[1] Taylor, Tennent and Whitaker, 1926 and 1929.
[2] Harnly (1926); cf. Hörstadius, 1937 a.

section has now been performed on *Arbacia* eggs, taking the micropyle as a guide for orientation, while other sections were made as far away as possible from the nucleus.[1] There is not the slightest doubt that the juxtanuclear material has no specific importance for the organization of the pluteus. It is equally evident that the vegetative part of the egg has potencies that are entirely missing in the animal half. This develops into larvae of animal type, while the vegetative part forms micromeres antipolar to the cut side, and later builds up primary mesenchyme, skeleton, and, naturally, entoblast. A critical revision of the above-mentioned data concerning *Lytechinus* at least throws a doubt on the validity of the equipotency of the territories, which was assumed in this species. The conclusion drawn for *Arbacia*, "that the micromere-forming and the entoderm- and the skeleton-forming material is located in the most vegetative part of the volume of the egg"[2] is probably valid for any sea-urchin egg.

Let us come back, for a while, to the operations where **micromeres are grafted** at various levels of a young morula. The demonstration of an inducing power is evidently of interest for comparison with Vertebrates, and there is no doubt that, in certain conditions, the grafted micromeres provoke a local invagination, which develops into a variable archenteron, surrounded with the primary mesenchyme proceeding from the grafted material. The use of the term "Organizer" is, however, not quite advisable. It might suggest that the micromeres are indispensable to the normal development, and that is certainly not true. On the other hand, certain organs of Vertebrates are known to possess an inducing power which has no function in normal ontogenesis, for instance, the neural tube.[3] It seems thus more cautious to avoid any generalization of the disputable term "Organizer", and to pay more attention to the conditions of induction. They have been recently summed up by Hörstadius in such a clear manner that I cannot refrain from quoting his authoritative description: "The gradients can also be illustrated

[1] Cf. Hörstadius, 1937 *a*.   [2] Hörstadius, 1937 *a*, p. 315.
[3] Objections formulated by v. Ubisch, 1936.

in the following way. If we implant four micromeres in the animal pole of an entire egg, we get only a very small archenteron induced. If we put the micromeres between An_1 and An_2, the archenteron will become larger, bipartite, and there will be supplementary skeletons on the sides. If we place the micromeres between An_2 and the vegetative half, a still larger archenteron will invaginate at the point of implantation, but this will later fuse with the normal archenteron.—Another experiment showing the struggle between the gradients involves the removal of the micromeres from the vegetative pole and their implantation into the animal pole of the same egg. We have here added nothing to the egg, only translocated the micromeres from one place to another, resulting in an enlarged digestive tract and a considerably enlarged coelom. In this case, the implanted micromeres do not induce an archenteron but they weaken the animal gradient so as to allow the vegetative gradient to express itself further towards the animal pole. The same explanation holds for some other cases when four micromeres are implanted into the animal pole of an animal half. In this experiment we sometimes obtain larvae with two archentera, one induced at the animal pole by the micromeres, one formed at the most vegetative part of the animal half. As isolated animal halves never gastrulate, this invagination of the vegetative part of the animal half seems also to be due to a weakening of the animal gradient by the micromeres. The most animal region of these larvae is situated between these two vegetative centres; here we have the apical tuft. In other cases there may be no invagination at the vegetative side of the animal half, but only at the point of implantation. A skeleton and stomodoeum are differentiated in accordance with this animal archenteron and the apical tuft may be developed at the most vegetative part of the animal half, which now constitutes the new animal pole. This means that we have in this animal half a complete reversal of the polarity of the egg axis brought about by the micromeres."[1]

Up to this point, we have carefully separated the results of **micrurgical and chemical methods** But both techniques may

[1] Hörstadius, 1936, p. 236.

be combined most successfully. Lithium has been used as a corrector of the animal tendencies inherent in the mesomeres. Like the graft of micromeres, it allows the production of normal plutei.[1] The same experiment may be performed in still more advanced stages, and then it shows, exactly like the serial implantation of micromeres, the gradual stabilization of the animal potencies. An animal half which has been allowed to develop some hours before being submitted to lithium fails to respond to its specific action, and that in a definite order: invagination of the archenteron is first inhibited, then the arms are no longer formed, finally, the stomodoeum and the ciliary band are affected. In a word, various territories of the animal vesicle resist the vegetativization more and more. This similitude of action between Li^+ and the micromeres made possible the demonstration of induction by "pseudo-micromeres". The animal half of an egg was isolated and submitted to Li^+; then, before any invagination of mesenchyme, the most vegetative part of that blastula was removed and placed against the equatorial region of an animal half taken from an untreated egg; instead of swelling, like the normal isolated animal halves of the same batch, into a hyperciliated blastula, the grafted system formed a normal pluteus.[2] With the accurate realization of the various necessary controls, this experiment is one of the finest technical performances in causal embryology.

The analysis of the intoxication provoked by the lack of sulphate has also taken advantage of the influence of Li^+ on the animal halves. These are not, by themselves, susceptible to the absence of SO_4^-. But, if they have been previously submitted to Li^+, and then deprived of SO_4^-, their entoblastic material shows the characteristic yellow and cloudy aspect.[3] The alteration is consequently not special to the subequatorial region of the egg, but is connected with the vegetative type of metabolism.

The animalization may be submitted to similar experiments. If vegetative parts of eggs, previously submitted to animalizing conditions, are isolated, they form normal plutei instead of defective ones, if the strength of the animalization has been suitably adjusted.

[1] v. Ubisch, 1929. [2] Hörstadius, 1936 *b*, p. 64.
[3] Lindahl and Stordal, 1937.

146   FIELD AND GRADIENT

All these tallying experiments justify, without any doubt, the conception that morphogenesis, in the sea-urchin, depends on the equilibrated interaction between two distinct and, up to a certain point, "hostile" activities.[1] We have now to go thoroughly into the **meaning** of these remarkable results. In this case, contrary to what happens for Vertebrates, each of these dominant tendencies is capable of an independent expression, giving rise to abortive structures of characteristic appearance and differentiation. Each of them is also linked to a special metabolism, the characteristics of which are at least partially known. The material substrate of these important functions probably has a general distribution corresponding to the two gradients involved. As regards its relation with the surface of the egg, a few remarks may be made. In *Strongylocentrotus pulcherrimus*, a cytological difference exists between the original cortical layer, crowded with pigment granules, and the part of the superficial film formed at each division, to cover the internal surface of the blastomeres.[2] Moreover, the well-known ineffectiveness of centrifugalization on the animal-vegetative polarity of the sea-urchin egg excludes the existence of some internal more or less rigid structure and points to the importance of the cortical layer, which precisely is not affected by centrifugalization.[3] Thirdly, the primary polarity of the egg, even if not irreversible (cf. effect of chemical conditions, graft of micromeres in the animal half), is nevertheless remarkably stable. When the two first blastomeres are separated and then united again with an alteration in their respective orientation, a unitary larva is only obtained when the axes remain parallel.[4] Fusion of the 2-cell stages only produces a unitary giant pluteus when the two axes are in prolongation of each other.[5] Inversion of the vegetative half of the 16-cell stage permits some rearrangement of the material, but the complete individuality is not restored.[6]

Besides these immediate statements, an interesting suggestion emerges from the comparison between the effects of Li+ in Amphibians and Echinoderms. In the first group, this intoxica-

[1] Hörstadius, 1936 c.
[2] Motomura, 1933; Dan, Yanagita and Sagiyama, 1937.
[3] Howard, 1932; Motomura, 1935; Lindahl, 1932, 1936.
[4] Peter, 1931.   [5] Balinsky, 1932.
[6] Hörstadius, 1928.

tion applied before gastrulation has two consequences:[1] in the marginal zone a uniformization takes place and the chorda material is transformed into somites; and a reduction of head-induction results in monorhiny, cyclopia and microcephaly. Both facts are easily interpreted from the theoretical viewpoint of the interaction between yolk gradient and dorso-ventral field, if a damaging of the second factor by Li^+ is assumed. The quantitative diminution of C (cf. p. 91) causes the ratio C/V to fall below the chordo-mesoblastic threshold, and the whole material develops into somites. On the other hand, the concentration of the $C.V$ products is reduced in head-inducing material, and a deficiency is observed in the two most susceptible sense organs and in the volume of the brain. Now, Li^+ is known, in the sea-urchin, to act upon the metabolic and morphogenetic activities of the animal gradient. On the other hand, it has been shown in *Paracentrotus lividus*, by ultramicroscopic observation, that fertilization results in the formation of a subequatorial yellow belt, approximately covering Vg_1, and that, under the action of Li^+, this yellow band extends towards the animal pole. Its enlargement has been noted as early as the 32-cell stage.[2] This fact indicates that the Li^+ intoxication brings about a perceptible modification of the cortex. Considering the whole situation, it has been proposed[3] to imagine that the animal activity of the Echinoderm egg is carried by a field similar to that suggested as the substrate of dorso-ventral polarity in Vertebrates. This suggestion does not exclude the participation of internal cytoplasm in animal activity; it only assumes that the condition which governs its intensity and the orientation of its reactions has a cortical localization. *We also consider the animal field as stable, providing the environment remains normal.*

The vegetative gradient has also evident properties in common with the yolk gradient of Chordates. Its apex has the same position and it marks the region that will invaginate first. A relation with cytoplasmic inclusions is indirectly inferred from experiments relative to dorso-ventral polarity (cf. p. 155). There is, however, no clear indication for either a cortical or an internal

[1] Lehmann, 1937.
[2] Runnström, 1928, p. 568.
[3] Dalcq and Pasteels, 1937.

localization. We will leave both possibilities open, and only *assume this gradient to be easily modifiable*. True homology with the Chordate yolk gradient is also excluded by the fact that the latter, as such, has no morphogenetic power, but only acquires it by interaction with the dorso-ventral field, while the vegetative gradient of the sea-urchin has a specialized capacity for organization.

Let us now consider how the field An and the gradient Vg interact in normal development and in experimental conditions. The number of questions to be solved is not less than that of the problems raised by the study of Vertebrates, but the solution that has been found there may be tentatively taken as a guide. A preponderant rôle has been attributed in that case to the product $C.V$. The immediate and necessary interaction of An and Vg is, however, not suitable for the present case. The autonomy of both the fundamental activities in experimental material indicates that the substrates of An and Vg are each able to react with ordinary components of the cell and thereby give rise to definite structures. Moreover, if the interaction $An.Vg$ was concerned from the beginning, the combination of their maximal values would have the most intense effects: and this does not happen, as is shown in the graft of supplementary micromeres in the animal region. The general conclusion reached from the researches described above is that the two morphogenetic activities primitively act by counterbalancing each other. " One of the main principles emerging from this work ", Hörstadius writes, "is that the development of a typical larva out of a fragment is due, not to the absolute amount of animal or vegetative material present, but to their relative amount."[1] This induces us to consider the rôle of the ratio Vg/An, a representation of the quantitative relation between the products of the two distinct activities.

The first morphological event resulting from these is gastrulation. That this involves an equilibrium between An and Vg becomes evident when the two opposite tendencies are considered in relation to the auxetic surface process. The whole of animalization is characterized by the preponderance of the external surface

---

[1] Hörstadius, 1936, p. 237.

growth over the internal, the extreme manifestation of this
difference being the generalized hyper-ciliation. From this point
of view, events happen as if the substance of the animal field
lowered the surface tension at surfaces of contact with the
medium. The vegetative activity, on the contrary, points to a
predominance of the internal surface growth on the external one.
In a normal egg, the maximum tendency is exhibited by the
micromeres, which quickly leave the continuous epithelium of
the blastula and expand in the blastocoele; after invagination
of the archenteron, a similar process is repeated for the secondary
mesenchyme. In abnormal conditions, where Vg is exaggerated,
the difference between the rates of growth on the two surfaces is
modified in the apical region of the gradient, deformation starts
outwards instead of inwards, but the essence of the process is
not altered. In short, there is an intimate relation between
the animal field and epiboly on the one side, and between the
vegetative gradient and invagination on the other side.

After this consideration of the ratio Vg/An, the concept of
threshold may be tentatively introduced. If we represent (fig. 54)
the normal variation of the ratio by the arbitrary curve α, we
may suppose that invagination occurs for any part of the egg
attaining a value of Vg/An superior to IL, which is the threshold
of invagination exactly attained at the boundary between Vg_1
and Vg_2. Let us now examine some typical experimental cases.
The suppression of the animal half does not move the limit of
invagination: the Vg gradient has not been disturbed, for its
apical part is intact, and the An field is only reduced far from the
limit of invagination. When Vg_2 is isolated, the limit of invagination is moved downwards: the field is, and remains, feeble, but
the gradient concentrates in the vegetative direction, the threshold being attained at about half the height of Vg_2. When the
micromeres are preserved together with Vg_2, exogastrulation is
observed: the threshold is attained at the animal apex of the
system. In operations where the vegetative part is reduced to
a sufficient but not too marked degree, the limit of invagination
moves proportionally upwards: the An field does not change, but
the ratio takes the form of the curve β; the limit of invagination
is displaced animalwards. When the amount of the vegetative
material is too far reduced, no invagination occurs: the excessive

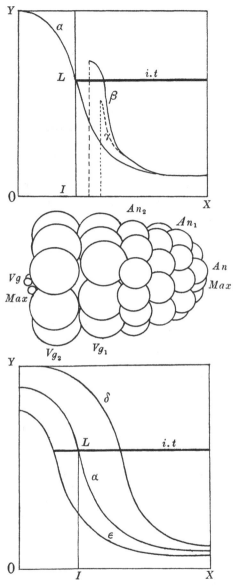

Fig. 54. Graphic representation of the quantitative conditions governing the determination of the limit of investigation. Above, approximative variation of the ratio Vg/An in a complete egg (α), and in cases of medium (β) and excessive (γ) reduction of the vegetative part. Beneath, modification of the curve α in case of vegetativization (δ) and of animalization (ϵ). LI, Ordinate traced in correspondence with the limit between Vg_1 and Vg_2 and giving the arbitrary value of the invagination threshold ($i.t$).

steepening of the gradient causes the curve to begin below the threshold value.[1]

The various effects of chemical conditions also easily find their expression in this representation. By its alteration of the cortical field and possibly its exaggeration of the Vg metabolism, lithium causes an elevation of the whole curve (fig. 54, δ) and the threshold of invagination is attained by more animal regions. The inverse holds good for unipolar animalization. The curious bipolar animalization implies that the treatment causes, in some eggs, a thorough change of structure in the animal field; this process needs, it seems, further investigation. The possibility of micrurgical and chemical combinations between animalization and vegetativization does not raise any difficulty.

The grafts of micromeres (p. 143) have different efficiencies according to the level at which they are placed, and transferring the micromeres of an egg from its vegetative to its animal pole is not equivalent to the addition of foreign ones. These two data are now intelligible. To cause an induction, in the sea-urchin, is to provoke a secondary and localized gastrulation. The ratio Vg/An must therefore be increased over the invagination threshold. This happens by the diffusion of an excess of vegetative substance from the micromeres, and it is necessarily easier for lower regions, with a primitively larger ratio. In the mere translocation of micromeres to the animal pole, the whole Vg gradient is disturbed and the substances escaping from the graft reinforce it in its animal part; the result is a shifting of the limit of invagination animalwards, with the resulting enlargement of the digestive tract and of the coelome, but a secondary induction necessarily fails. These few instances are sufficient to illustrate the efficiency of the representation now suggested, on the basis of a quantitative appreciation of the two chemical systems which endow the Echinoderm egg-cell with the quality of a germ.

We have now to consider the gradual establishing of **differences between the ventral and dorsal regions** of the germ. The simplest

---

[1] It has been noted (Hörstadius and Wolsky, 1936, p. 97) that the complex $8 + Vg_1$ has distinctly more vegetative activity than the four meridian quarters of the same complex; they yield merely hyperciliated blastulae. It seems possible that the originally elongated shape of these quarters flattens their Vg gradient more than is the case in the whole group.

method of studying this process consists in obtaining frontal, sagittal and oblique sections of the egg. The late appearance of the indices of bilateral symmetry is no hindrance to these operations; preliminary vital staining makes it possible to ascertain the orientation of the cut after the experiment is over.

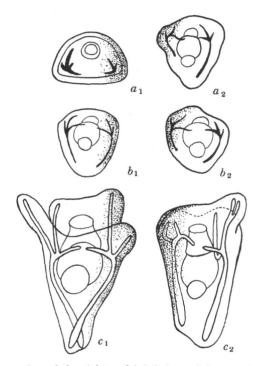

Fig. 55. Separation of the right and left halves of the morula. a_1, b_1, c_1, three successive stages of the left half, with the retardation of the internal, vitally coloured (stippled) side. a_2, b_2, c_2, mirrored forms furnished by the right half. Redrawn from Hörstadius and Wolsky, 1936.

The plane of cleavage of the first two blastomeres does not present any fixed relation with the plane of symmetry or with the point of fertilization. If the furrow has been coloured, it will be easy to recognize the cases corresponding to either a sagittal, a frontal, or an oblique section, especially if the separation has been formed at the 16-cell stages. In the first case, the larvae generally[1] exhibit a retardation of the stained region, which is primitively internal; the growth of the arm processes and the formation of the skeleton are distinctly slower and the embryos are remarkably asymmetrical (fig. 55). In cases where the section has been frontal, two important facts are noticed: the dorso-ventral axis of the dorsal half is inverted; its development is clearly retarded (fig. 56). Oblique sections give intermediate results.

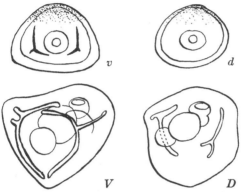

Fig. 56. Separation of the ventral and dorsal halves of the fertilized egg. v, V, The ventral half, with the coloured area (stippled), indicating the surface on the dorsal region. d, D, The dorsal half, also coloured on the dorsal side (inversion !), and distinctly inhibited in its development. Redrawn from Hörstadius and Wolsky, 1936.

These results may be generalized to operations made on younger stages, certainly up to the 2-cell stage, and probably hold for the unfertilized egg. In this latter case, batches with a distinct red pigment girdle must be selected and median sections made at random. A certain number of the twin-embryos resulting from such double merogony show a mirrored symmetry (fig. 57, a, b, c). As a counterpart of it, certain pairs exhibit, in each member, the same curvature of the archenteron in regard to the calcareous spicules (fig. 58), and seem to indicate that a difference between the dorsal and the ventral regions exists before fertilization. The elongation of the egg perpendicularly to its axis in certain

[1] In 23 out of 28 cases. Hörstadius and Wolsky, 1936, p. 77.

154 FIELD AND GRADIENT

Echinoderms (*Holothuria, Asterina gibbosa*) is direct evidence confirming such organization.

When the same operations of meridian section are performed on still more advanced stages, the internal deficiency becomes more and more evident (fig. 59). With sagittal sections, young gastrulae yield typical hemiembryos. With frontal divisions, the dorsal inversion of the axis is observed until the migration of the

(a) (b) (c)

Fig. 57. Three pairs of left-right halves, obtained by operation on the unfertilized egg. The difference of stage is due to the haploidy of one member in each pair. From Hörstadius and Wolsky, 1936.

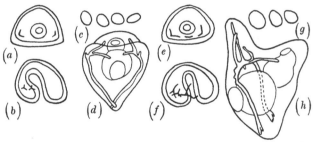

Fig. 58. A pair of larvae obtained by median section of the unfertilized egg. (a), (b), (d) Three stages of the haploid member, with its nuclei, (c); (e), (f), (h) Idem for the diploid member, with its nuclei, (g); the haploid larva seems to have arisen from the ventral half of the egg. Redrawn from Hörstadius and Wolsky, 1936.

primary mesenchyme. Once the gastrulation has started, the plutei become deficient respectively on their dorsal and their ventral regions, which means that the inversion can no longer take place. In the intermediate period, the primary dorsal half is sometimes found to give rise to a pluteus with two ventral regions.

This susceptibility of the dorso-ventral axis to experimental conditions is also to be observed in the entire egg.[1] When it is

[1] Boveri, 1901; Lindahl, 1932 *a* and *b*.

forced, before fertilization, through a narrow pipette, it may happen to be elongated perpendicularly to its primary axis, and the acquired shape persists for some time after the expulsion from the pipette. If the modified egg is fertilized, its long axis is maintained as the dorso-ventral axis of the pluteus. Vital colorations allow the experimenter to assert that the part of the egg which penetrated ahead into the pipette generally corresponds to the induced ventral size. The phenomenon is, however, not due to the change of shape in itself, but more probably to a slight lesion, modifying the colloidal structure. In this way, a new

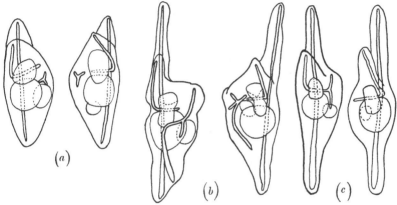

Fig. 59. Results of the sagittal section of the sea-urchin egg at middle stages. (a) A pair from an egg cut 10 hours after fertilization. (b) The same embryos at an older stage. (c) Isolation performed 16 hours after fertilization. From Hörstadius, 1936.

ventral centre is substituted for the one presumed to exist in the normal unfertilized egg. Centrifugalization before fertilization also influences the dorso-ventral axis. In *Psammechinus*, the centripetal region of the egg, where fat globules accumulate, regularly becomes ventral. In *Arbacia*, this fate awaits the centrifugal part, where yolk inclusions are condensed. The sole common point in these two experiments seems to be that ventral structures form where a certain amount of nutritive material has been condensed and only dissolved in later stages.[1] There is some analogy between these experimental conditions and the displace-

[1] Lindahl, 1932 b.

ment of yolk in Amphibians. The determination of the ventral region appears to be correlated with some preponderance of the vegetative activity. It may now be ascertained that this situation already exists before cleavage. When fertilized but undivided eggs are submitted to KCN, a special deformation appears, a sort of plasmolysis with a deeper hole on one side. The spherical shape is gradually restored during cleavage. Vital marks show that the most sunken part corresponds to the ventral region of the larva.[1] This is therefore characterized, from the very beginning of development, by a particular physical and probably chemical constitution.

The results just described concerning the gradual appearance of the bilateral symmetry, with dorso-ventral polarity, are dominated by two main facts: certain chronological differences appear between the regions of dwarf-embryos or between the members of a pair; and the dorso-ventral polarity of dorsal halves is inverted by isolation. The retardation of parts or regions calls to mind the chronological difference in the double gastrulation of inverted frog's eggs. In both cases, it suggests the rôle of a chemical reaction producing active substances. In the frog's egg, the useful consideration of the product $C.V$ started from this fact. In the sea-urchin's egg, it indicates that, beside the ratio Vg/An which controls invagination, the positive interaction $An.Vg$ must also be involved in the interpretation. We have previously assumed that the animal cortical field is endowed with a considerable stability. The planes of contact between the blastomeres only possess this cortical substance in the amount drawn inside by the formation of the furrows, and cytological observations (p. 146) indicate that this extension in the depth is probably slight. A deficiency of the surface of previous contact, in median sagittal sections, will necessarily reduce the amount of $An.Vg$. This accounts for the retardation, but, on the other hand, the very fact that differentiation takes place, although delayed, teaches us that the $An.Vg$ substances have a certain power of diffusion, exactly as in Vertebrates.

Turning now to the dorso-ventral sections, we are induced by the results of stretching and of centrifugalization, by the inflexion of the archenteron, by the special activity of the ectoblast in the

[1] Foerster and Örström, 1933.

region of the arm processes,[1] to postulate a larger concentration of the $An \cdot Vg$ substances on the ventral side of the primary axis. This view finds its confirmation in the rate of development observed in ventral and dorsal halves; it is faster in the former (fig. 56). The reversal of the dorso-ventral axis in dorsal isolated fragments will now find an easy explanation.[2] From our present viewpoint, the concentration of the $Vg \cdot An$ substances is higher in the ventral region. This may be due to a shifting of the gradient apex relatively to the axis,[3] or to a slower decrease of intensity on one side, or to an easier diffusion of substances produced near the vegetative pole.[4] The essential point, on which all students of the question seem to agree, is a predominance of the vegetative activity. This means that a secondary cortical field results from the $An \cdot Vg$ reaction, with a focus placed somewhere on the ventral meridian (fig. 60). Now, when the young dorsal and ventral halves are separated, the deficiency of the cortical field on the newly acquired surface affects the distribution of the $An \cdot Vg$ potentials. In the ventral half, the decrease of the $An \cdot Vg$ field is exaggerated, but probably amended gradually by diffusion. In the dorsal half, the normal situation is inverted and, in spite of the later diffusion, the focus of the field is brought on to the primitively dorsal side and this is bound to become ventral (fig. 60, b, c).[5]

By this last problem[6] we have been led to take into consideration the secondary appearance, in the sea-urchin development,

[1] The fundamental stimulus to the disposition of the skeleton starts from the ectoblast, which determines the arrangement of the primary mesenchyme. The skeleton pieces then influence the growth of the arm processes. Through this reciprocal action, the location of the arms remains essentially governed by the ectoblast. (Runnström, 1929)

[2] Dalcq and Pasteels, 1937.

[3] Schleip, 1929; Runnström, 1931; Lindahl, 1932.

[4] Lindahl, 1932. This author also admits (1936) an easier diffusion of animal substances.

[5] An attempt to explain the same process has also been made by Bernstein, in the discussion of the interesting symposium held last year in Woods Hole (Hörstadius, 1936 a). Admitting a decreasing gradient of a substance from the ventral to the dorsal side, it is suggested that the contact of sea-water on the surfaces of sections causes this substance to escape. The primitive gradient is thereby cut into two opposite ones.

[6] For an accurate discussion of the intricate problems of metabolism related to the dorso-ventral polarity, the remarkable thesis of Lindahl is to be referred to.

of chemical reactions between substances related to the two distinct metabolisms. This is a most logical assumption. In Echinoderms, as in Chordates, development runs towards a **mosaic of determined regions,** through an intermediate phase in which more and more limited secondary fields become superposed on the general morphogenetic pattern. The introduction of the symbol $Vg.An$, together with the fundamental influence of the ratio Vg/An, and with the general importance of specific thresholds, accounts for the organogenetic interactions between

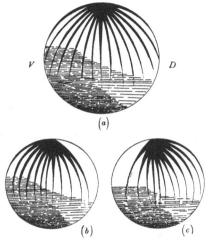

Fig. 60. Diagrammatic explanation of the reversal of the dorso-ventral axis in the isolated dorsal halves. (a) The entire egg, with its animal field and vegetative gradient. (b) and (c) The ventral and dorsal halves respectively. D, dorsal; V, ventral.

the three layers established by gastrulation: the delineation of the ventral area in the ectoblast, its influence on the archenteron and on the primary mesenchyme; the localization of the stomodoeum; the action of the skeleton pieces on the growth of the arms, etc. The final determination is then attributed, as in Chordates, to the definitive interaction between the substances produced by the morphogenetic metabolism and the ordinary constituents of the cell.

The present review of the morphogenetic organization of the

sea-urchin's egg is unavoidably very incomplete. Limited as it is to the most conspicuous processes of early development, it shows the meaning of the law that Driesch formulated with such extraordinary prescience: the fate of each cell is a function, mathematically speaking, of its position in the whole. But the main advantage of the present survey is the possibility of an exact **comparison** between the germinal organizations of Chordates and Echinoderms. In both instances, two main elements are responsible for morphogenesis. In both cases, one is a cortical field, the other a spatial gradient. In both groups, they act by their own activity according to their ratio, and by their chemical interaction according to their products. By these fundamental processes segregation gradually takes place inside the primitively continuous system, on account of definite thresholds. There are, however, essential differences in the substrates of organization, as well as in the morphogenetic physiology. In Chordates, field and gradient have, as far as is yet known, no autonomous action, while this is quite conspicuous in the sea-urchin. In the former, the first steps of segregation are caused by the interaction of field and gradient, their relative antagonism acting secondarily. In the latter, the ratio controls the first events, while the products of reaction afford the secondary complications. In a word, there is identity in the principle, but difference in the application. Nothing else could be expected for so distantly related species.[1]

[1] *Addenda.* Ranzi and Falkenheim (1937, 1938), using various vital staining dyes, conclude that the animal material has a lower rH than the vegetative region. The vegetativizing effect of Li is accompanied by an elevation of the rH values. (Publ. Staz. Zool. Napoli, **26**, p. 436; *Naturwis.*, **26**, p. 44.)
The interpretation given pp. 156–157 and the representation of fig. 60 should be improved by admitting, with Lindahl, that the presumptive ventral side is characterised by a slower decrement of *both An* and *Vg* gradients. The objection would thereby be avoided, on which Dr Hörstadius has kindly called my attention, that the course of gastrulation is typically monaxone, which should not be logical if vegetative forces only were prevalent on the ventral side.

CHAPTER X

Regulation in some other eggs

In the preceding pages, we have dealt at some length, but not exhaustively, with the early morphogenetic processes in Prochordates, Vertebrates and Echinoderms. The special attention paid to these three classes is justified by the importance of the results already obtained and by the intimate correlation between the conclusions arrived at. Extensive investigations have also been made on representatives of most other classes and must be considered if we wish for a panoramic view of the information available. But, except for one or two quite recent researches, they have already been fully taken into account in the valuable books mentioned on p. 4, and we may limit ourselves to a brief outline of the experiments where regulation is concerned.

In **Coelenterates,** we have precise evidence about Hydroids, some rather ancient observations concerning hypogenetic Medusae, and a large amount of study on the Ctenophore egg. From his experiments on Hydroids, Maas has traced the scheme of an egg containing a vacuolated, entoderm-forming entoplasm, surrounded by a dense, ectoderm-forming ectoplasm. Groups of blastomeres equivalent to a half or a quarter of an egg often develop into a normal polyp, thanks to the extension of the ectoplasm over their whole surface. But as soon as the cells fall below a certain size, their potency becomes limited. When the arrangement of the blastomeres is completely disturbed by stretching them in a chain, development may be normal, a fact that is hardly compatible with any complex internal organization.

Teissier's researches led to very considerable advance in this direction. Two of his species, *Hydractinia* and *Dynamena*, have eggs of the preceding type. Their regulative power is great, especially in the blastula. It is interesting to note, as introducing a quantitative point of view, that a partial blastula of *Dynamena* only produces a symmetrical colony when above a certain size. According to the amount of the embryonic tissues, the germ builds up one or two hydranths. In a third species, *Amphisbetia,* the polarity is very conspicuous. An orange-coloured pigment marks the pole of maturation (fig. 61) and the upper blastomeres. The blastula elongates along the axis through this spot, and swims with its white pole forward, while at the opposite region

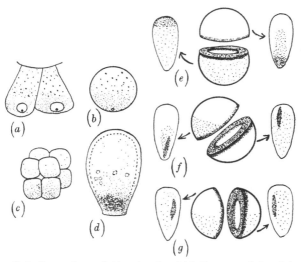

Fig. 61. Polarity and regulation in *Amphisbetia operculata.* (a), (b), (c) Ovarian oocyte, mature egg and 8-cell stage, each with the pigment granules at the maturation pole; (d) swimming blastula with its pigmented anterior pole. (e),(f),(g) Sections of the blastula, with coloration of the cut surface; left and right, the planulae formed by each part; the primitive axis is maintained both in either equatorial (e), oblique (f) or meridian section. Redrawn from G. Teissier, 1931.

entoblastic cells migrate inside, one by one. Gastrulation having taken place in this way at the orange-coloured maturation pole, the planula becomes fixed, after four or five days of free existence, by its anterior white part. The polarity of the egg can thus be followed without interruption from the oocyte stage. The equatorial third furrow isolates four uncoloured, ectoblastic cells from four pigmented, entoblast-forming ones. Vital staining confirms this repartition of the anlagen. The same method is then combined with the isolation of the blastomeres at the 2-4-6-8-16-cell stages, and of parts of the preblastula as well as of the blastula. All these portions of the germ form a planula of reduced size, but of normal structure, without any change of polarity (fig. 61). The stability of the primary axis is confirmed by experiments in which blastulae are joined together, without ever fusing integrally into one individual, and by displacements of blastomeres, where groups of distinct polarity are often to be recognized. Some normal planulae are however obtained, but they may result from a re-establishment, by chance, of the concordance between the axes. Polarity is thus the sole stable feature of the reduced or complex systems studied in *Amphisbetia*: the fate of the cells varies according to their situation along the animal-vegetative axis.

This result affords a most useful basis for comparison between the eggs of Hydroids, Medusae and many Coelomates. Eggs of Medusae generally present a more complex organization. Some simply add to the polarization a gelatinous mass somewhere on the main axis, and this regional structure affords, in *Geryonia* and *Lyriope*, a limitation of the power of regulation.[1] With Ctenophores, the specialization is quite remarkable. The unfertilized egg consists of a mass of loose entoplasm surrounded with hyaline ectoplasm. In *Beroe*, this external layer assumes, under the ultramicroscope, an emerald-green colour, and absorbs the vital staining dyes,[2] not on account of a pH difference but of metachromasy.[3] This material alternatively undergoes a condensation and an expansion following the rhythm of cleavage. After the third division, the micromeres, which are formed opposite to the maturation pole, receive nearly the whole "green" substance, which now remains localized. It is divided between the daughter-cells of the micromeres, disposed in a ring surrounding the so-called "animal" pole. The small portion that has been left in the macromeres is soon separated from them during an unequal division and passes into the "vegetative" micromeres, which will take part in the formation of the digestive wall. The largest portion of the emerald plasm thus becomes included in the ectoblast, just where the ciliated bands will later appear. These ribs, when observed with the ultramicroscope, are nevertheless

[1] Maas, 1905; Schleip, 1929.   [2] Spek, 1926.   [3] Teissier, 1934.

deprived of any special coloration. This recently described develop-ment[1] explains the results[2] previously obtained in experiments where the egg was cut, or its blastomeres separated, displaced, translocated, etc. The presence of the specialized plasm is necessary for the differentiation of the ciliated bands, a process still obscure in its details. Operations on the undivided eggs have effects in so far as they affect the distribution of the emerald substance. The precocious excisions may be, according to the stage, either compensated or followed by deficiencies.[3] But their interpretation, as an evidence of a precocious localization in certain parts of the cytoplasm of factors corresponding to each of the ciliated bands, is certainly an error. After separating the first blastomeres, their cleavage progresses normally and nothing is changed in their fate, except some regulative process concerning, in hemi-embryos, the proportions of ectoblast and entoblast, and also the statocyst. In short, early determination is, in this case, intimately bound with cell division, and the displacements of the emerald plasm appear to be passive, caused by variations of surface tension during plasmodieresis. The orientation and position of the mitotic spindles would thus represent a primary morpho-genetic factor. It is however not excluded, in my opinion, that the physical properties of the cortical material, its own surface tension, and the pressure it exerts on cytoplasm during anaphase, could be responsible for the quite unequal plasmodiereses observed after the 8-cell stage. From the preceding facts, it appears that double merogony, if it could be performed on the Ctenophore egg, would probably reveal an extensive regulation. From the study of the undivided egg and of its blastomeres, the rôle of the cortical substance is clearly demonstrated. We have to note its abundance, its rather long persistence on the surface, and the disappearance of its characteristic appearance at the moment when it becomes integrated in the differentiated material: this means that a reaction has occurred with some inner cell com-ponent.

The various classes generally designated by the embryologists, according to a common type of cleavage, as **Spiralia**—Annelids,

[1] Spek, 1926.   [2] Driesch, Chun, Morgan, Fischel, Yatsu.
[3] Fischel, 1903.

Hirudinates, Nemertines, Scaphopodes, Gastropodes and Lamelli-branches—have for long been held to be lacking in any regulative power. Their extremely precise cell lineage seemed to be an expression of a rigid determination. Thanks to progressive analysis, this illusion has now vanished. In general, the structure of the egg shows certain differentiations: an ectoplasm, sometimes incomplete (*Chaetopterus*); a vegetative pole-plasm fairly constantly; and an animal pole-plasm in addition in Oligochaetes and Hirudinates. Without entering here into a detailed account of these highly specialized developments, we shall record that, in many species, the *D*-quadrant of the 4-cell stage is remarkable for its volume. It includes the whole vegetative pole-plasm. When there is a distinct animal differentiation, this plasm also becomes included in the *D*-cell, where both specialized materials leave the cortex and fuse against the nucleus. In cases where a polar lobe is formed during the first two divisions, as in *Dentalium* and *Chaetopterus*, its substance is also incorporated within the same *D*-blastomere. In all species, two of the *D*-daughter-cells will become the somatoblasts. Both are placed on the same meridian, one just over, the other just under the blastoporal lip, and the principal elements of the entoderm and of the ectoderm will proceed from these cells. The first somatoblast (2*d*) forms the somatic plate, the second (4*d*) is the origin of the "meso-dermic teloblasts", the rôle of which cannot be described in this rapid outline, limited as it is to the points concerning the manifestations of regulation.

In Nemertines, the researches of the pre-War period[1] had already attained substantial results. It had been shown that any part of the unfertilized egg was able to form an apparently normal pilidium; that half-blastomeres segment typically, as if the missing blastomere was present; that they nevertheless give rise to a pilidium which may be asymmetrical in some cases. Between fertilization and cleavage, a restriction of the potencies had been observed concerning the animal and vegetative parts of the egg. This happens gradually, so that the animal half of an egg which has already given off the polar bodies is still able to gastrulate and form a pluteus. Experiments performed on the 8- and 16-cell stages demonstrated, on the contrary, a strictly

[1] E. B. Wilson, Zeleny, Yatsu. Cf. Hörstadius, 1937 *b*.

partial development. Operations on the blastula seemed, however, to admit some larger power of regulation, animal pieces being able to gastrulate.

These indications have been recently tested by Hörstadius,[1] using the methods which have yielded him, on the sea-urchin egg, such consistent results. The potency of any sufficiently large part of the unfertilized egg to form a normal pilidium has been confirmed. Stages between fertilization and cleavage have not been reinvestigated. But the potencies of the blastomeres have been tested most accurately. The well-known structure of the pilidium-larva offers (fig. 62, *e*) many characteristic differentiations: a ciliated ectoblast with an apical sense organ, a long flagellum inserted in a thickened pit of the ectoblast; bordering the mouth, two lappets with a specialized ciliation; an oesophagus and stomach, ending in a cul-de-sac orientated dorsally. For the 2-cell stage, the main interest consists in the relation of the cleavage plane to the bilateral symmetry. Vital colorations indicate that the cleavage simply cuts from the animal to the vegetative pole without any correlation with the plane of symmetry. But the isolation of half-blastomeres (fig. 62, *f*) does not show clearly if the cut has been sagittal, frontal or oblique: the bilateral organization is not yet sufficiently established to get deficiencies on the side where material has been removed. The 4-cell stage is still formed of equal cells; each of these quarter-blastomeres forms a dwarf pilidium (fig. 62, g_1 to g_4). With the 8-cell stage, attained by a dexiotropic division, the first quartet of micromeres is formed at the animal pole, but they are so large that, as in the sea-urchin, the distinction between the animal and the vegetative halfs is easily seen (fig. 62, *c*). The boundary between these falls into the ciliated band bordering the lappets (fig. 62, *c*, *e*). The four animal cells, when isolated, do not gastrulate, but form a ciliated vesicle with an apical organ and, at the opposite side, a typical ciliated band (fig. 62, h_1). The only anomaly is that more than one apical organ is often present (fig. 62, h_2). But this may be tentatively explained by the assumption that all four cells contribute to the sense organ, and that their animal part has been somewhat shifted by the operation. If this is the case, the development of the animal half

[1] Hörstadius, 1937.

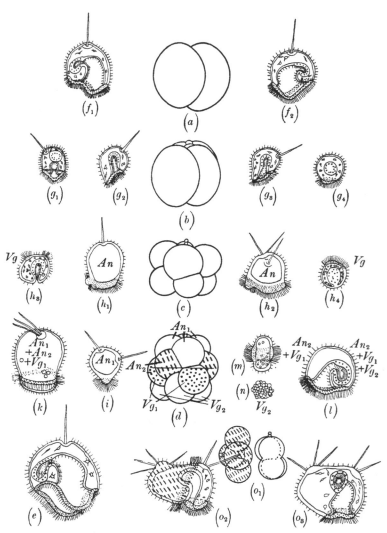

Fig. 62. Analysis of the potencies in the *Cerebratulus* egg. (a), (b), (c), (d) Stages with 2, 4, 8 and 16 blastomeres. (e) The normal pilidium. (f_1) and (f_2) Dwarf pilidia issued from half-blastomeres. (g_1), (g_2), (g_3), (g_4) Dwarf pilidia formed by quarter-blastomeres. (h_1) and (h_2) Partial larvae formed by the four animal blastomeres of (c). (h_3), (h_4) Idem by the four vegetative blastomeres. (i), (k), (l), (m), (n) Results from the dissociation and recombination of the four slices of (d). (o_1), (o_2), (o_3) Combination of a meridian half with an animal half, as shown in (o_1), with the formation of an "additive" embryo and its subsequent partial regulation. From Hörstadius, 1937b; modified.

would be a strict normogenesis. The same may be asserted concerning the four vegetative cells, with the important remark, that their material has succeeded in gastrulating (fig. 62, h_3, h_4). This means of course, in my opinion, a displacement, in the vegetative direction, of the limit of invagination. As a result, the ciliary differentiation, which normally occurs round the blastopore, is now shifted to the animal part of the partial larva.

When the 16-cell stage is reached (fig. 62, d) by a leiotropic division, the four groups of blastomeres may be conveniently designated, to avoid the complicated terminology of the micromeres and macromeres, as An_1, An_2, Vg_1, and Vg_2. The limit between the two first of these tiers seems to fall about the equator of the pilidium, while the separation between the two last ones is found somewhere in the oesophagus. Isolation of these groups of cells does not afford anything really unexpected except that An_1 sometimes forms two apical organs (fig. 62, l), and that Vg_2 simply gives a small mass of cells, without any morphogenetic activity (fig. 62, n) exactly as was the case for the sea-urchin micromeres. Isolations of $An_1 + An_2 + Vg_1$ (fig. 62, k), of $An_2 + Vg_1 + Vg_2$ (fig. 62, l), of $An_2 + Vg_1$ (fig. 62, m) also fall into the line of strict normogenesis. Two transplantation experiments have been performed. The combination of $An_1 + Vg_2$ does not preclude the regulative formation of an oesophagus. The fusion of an animal and a meridional fragment of the 8-cell stage (fig. 62, o_1) seems, in the young larva, merely additive (fig. 62, o_2), but when the larva is allowed to grow older, a redifferentiation occurs, with a clearly regulative character (fig. 62, o_3).

Although as yet less complete than their author wished them to be, these experiments throw great light on the organization of a Spiralia-egg, in a case where there is no plasmatic polar differentiation nor precocious inequality of the blastomeres, nor formation of a vegetative lobe. On the one side, the regulative power of the unfertilized and of the recently fertilized egg is larger than in the sea-urchin egg. On the other side, the restriction of potencies happens faster, from the beginning of cleavage and probably sooner. In both situations, the analysis is considerably hindered, but the results point to a probable similarity of organization. One fact, that seems to me typical, is the mode of development of the vegetative half at the 8-cell stage, with its dissociation between the level of gastrulation and the site of differentiation. It indicates that the forces producing gastrulation and differentiation are of a somewhat different nature. That is what we have assumed above for the sea-urchin, considering in one case the ratio Vg/An, in the other the product of their values. So that it does not seem unreasonable to think that an explanation of the Nemertine data could also be given by a gradient-field interaction, according to definite reaction thresholds. The late redifferentiation is perhaps an indication that the Vg. An products are able to be liberated when they are not normally correlated with the general pattern of the whole system, and that they are then distributed again along the vegetative axis of the artificially created complex.

In the Scaphopode *Dentalium*, the importance of a material located at the vegetative pole has been demonstrated both before and after fertilization.[1] Only meridian halves of the unfertilized egg yield, by double merogony, two normal trochophores. None of the blastomeres, neither *AB* nor *CD*, neither *A*, *B*, *C* nor even *D*, is really totipotent. It seems that a collaboration, perhaps quantitative, is necessary between these cells. The combination of various micromeres and macromeres would probably be worth while. The rôle of the vegetative specialized region varies in the course of cleavage. In the undivided egg it controls, together with the organogenesis of the post-trochal region, the differentiation of the apical organ, which does not appear in eggs of which the polar lobe has been removed. The same operation made at the second division has no longer the same effect, because the character "apical organ" is already determined in the animal part of *CD*. Recent investigation[2] has shown that this conclusion holds for the Polychaete *Sabellaria*. It seems, however, that in this species the factor "flagellum" is situated in the upper part of the vegetative pole-plasm. Included in the first polar lobe, that factor comes into *CD*, but it is not extruded in the smaller second lobe and remains in the *C*-blastomere. A representative of the Oligochaetes, *Tubifex*, possessing two opposite pole-plasms, has been submitted to a most accurate study.[3] By various methods, especially with ultraviolet puncture,

[1] E. B. Wilson, 1904.
[2] Hatt, 1932.   [3] Penners, 1926 et seq.

the blastomeres have been eliminated successively. In the absence of A, B and C, a normal embryo is possible, while the elimination of D limits the development to a formless mass of ectoderm and entoderm. It has been shown that, from the material of this big blastomere, only the parts belonging to $2d$ and $4d$ are indispensable for normal morphogenesis. Interesting features have been discovered by the destruction of the first somatoblast or certain of its daughter-cells. The resulting deficiency, apparently quite normogenetic, is not persistent. If the embryos are reared a sufficiently long time, the mesoblast forms, by a special regulation, the central nervous system (Penners, 1936). In a leech, *Clepsine*, the situation is nearly the same, but CD is necessary for a normal development.[1] In both the last cases, the substance of both plasms is thus demonstrated to be necessary; their suppression or their alteration cannot be compensated by other material. Their dislocation by centrifugalization also inhibits progressive organization. In *Clepsine*, direct modification could be made of each plasm separately. Injury of the animal pole stops the cleavage. A lesion of the vegetative region is not equally injurious, but the development of the somatoblasts seems to be prevented.

All these results agree in demonstrating the functional preponderance of the D-blastomere. Definitive evidence of its importance is provided by the fact that a simple equalization of the first cleavage causes the egg to develop into a double embryo. This effect has been obtained by various means in *Tubifex*,[2] *Nereis* and *Chaetopterus*[3] (fig. 63). In the last species, regulation is also observed in any type of merogony even if the fertilized parts are obtained by centrifugalization, and deprived of any granules other than the unseparable micromeres.[4] Thus there are various degrees of regulation for the Spiralia egg. Their capacity for this process does not seem limited in the unfertilized egg of *Chaeto-*

[1] Mori, 1932. [2] Penners, 1926.
[3] Tyler, 1930. I shall take this opportunity to mention the process of "differentiation without cleavage" exhibited by this species and some related ones (F. R. Lillie, 1902). The differentiation is remarkable for some general analogy with the normal larva; although no organ is of course formed, the cyto-differentiation (vacuoles, cilia) is rather extensive. The absence of cleavage is only a relative one. The nuclear activity is permanent and important (J. Pasteels, 1934; J. Brachet, 1937).
[4] E. B. Wilson, 1930.

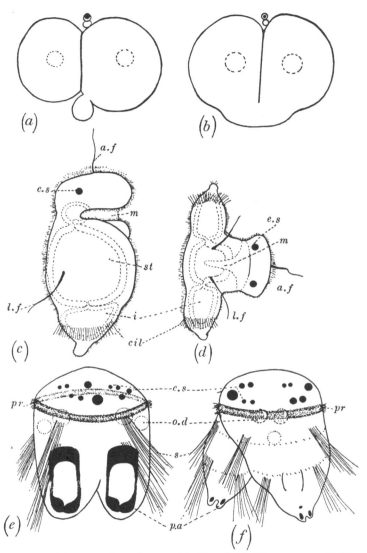

Fig. 63. Regulation in Polychaetes. (a), (b), (c), (d) *Chaetopterus*. (a) First normal division. (b) First equalized division. (c) Normal larva. (d) Double larva. (e), (f) Double embryos obtained by the same equalization of cleavage in Nereis. *a.f*, apical flagellum; *cil*, ciliary band; *e.s*, eye spot; *i*, intestine; *l.f*, lateral flagellum; *m*, mouth; *o.d*, oil drops; *p.a*, pigment area; *pr*, proto-chord; *st*, stomach; *s*, setae. Redrawn from Tyler, 1930.

pterus,[1] nor in the uncleaved egg of *Cerebratulus*, while in *Dentalium* a limiting condition, homologous to a vegetative pole-plasm, exists prior to fertilization. In the species where the first division is unequal, artificial equalization produces double embryos. If the first cleavage remains normal, the potency to form an entire embryo nevertheless still exists in a part of the egg, when isolated: the *C-D* group in *Clepsine*, the *D*-quadrant elsewhere. But this last point is not entirely valid for *Dentalium* or *Sabellaria*, and, on the other hand, the regulative ability of *CD* and *D* does not exclude the capacity of other blastomeres to develop, when isolated, as if they were left *in situ*.

A remarkable feature of these forms is that the initially unequal cleavage of many species is necessary to the formation of a single embryo. The deviation of the first cleavage plane effectively guarantees individuality. There is thus an intimate relation between cleavage and morphogenesis. This gives considerable interest to the elucidation of the conditions causing the unequal division. By accurate study[2] the conclusion has been reached that the eccentric position of the spindle is due to the presence of the vegetative pole-plasm, more or less concentrated in the eggs of various species. This cytoplasmic differentiation seems to be, by its physical state, inimical to plasmodieresis, and causes either the deviation of the furrow or the formation of the polar lobe. The physical properties of the substance playing the leading rôle in morphogenesis thus insure its utilization as a whole for the organization of one individual.[3]

In Nematodes, also, a deviation of the first cleavage is known to modify the potency of the first two blastomeres, as was shown by the celebrated experiment of Boveri and Hogue. But, even if the blastomeres issued from a sagittal division are equivalent, they are not thereby made totipotent. The atypically orientated mitosis has simply caused them to be homologous with the

---

[1] Recent observations of T. H. Morgan (1937) based on a close survey of various centrifugalization results, lead to the conclusion that "the egg of *Chaetopterus* has its dorso-ventral relation determined both before maturation and before fertilization".

[2] Pasteels, 1934.

[3] The question of the spiral cleavage is here entirely reserved. It is thought to be correlated with some intimate physical structure of the cytoplasm, which has been acquired under the influence of certain genes.

normal P-blastomere, characterized by its undiminished chromosomes. Like this cell, they are not able to form a whole embryo, and a complete regulation is not attained.

Finally, **Insects** must be considered. Until 1928, experiments on ants, beetles and flies had only revealed interesting facts of strict determination. In several species a special plasm, destined to form the initial germ-cells, had been demonstrated in the posterior region, but did not seem to concern morphogenesis.[1] When, however, the problem was attacked in other species, and by a scientist imbued with the analysing spirit of modern embryology, most interesting relations and co-ordinations were soon discovered.

About ten years ago, Seidel established that the egg of the dragon-fly possesses, at its posterior pole, a special plasm, which plays a leading rôle in the organization. It is not immediately active by itself, but only when some nuclei have been settled in its territory. From this nucleo-plasmatic reaction a substance results, the forward diffusion of which is indispensable to development of the embryo. This posterior differentiation of the egg cytoplasm is designated as the *formative centre*. On the other hand, normal features of ontogenesis also fix attention on the future thoracic region. This territory corresponds to the level of the ventral area where the nuclei first concentrate into an embryonic anlage. Admirably combined observations and experiments indicate that this activity is not connected with the production of a special substance, but with a particular reaction originating in the yolk plasma. The mass of deutoplasm is again something more, in Insects, than the purely nutritive reserve which it is generally thought to be. After its colonization by the so-called vitellophagous nuclei, it becomes capable of a sort of contraction, and this active process causes the migration and suitable arrangement of the embryonic nuclei in a definite area. The factors located there are indispensable to morphogenesis. They constitute the *differentiation centre*, the activity of which requires the previous diffusion of the substance elaborated in the formative centre.[2]

---

[1] Strindberg, and Hecht in *Camponotus*; Hegner in *Leptinotarsa*; Reith and Pauli in *Muscidae*.   [2] Seidel, in *Platycnemis* (1926 to 1936).

These fundamental conceptions have been the basis of new advances, and particularly of the discovery of regulation. In the honey-bee,[1] the thoracic centre is particularly clear. Ligaturing experiments show that determination is accomplished 24 hours after the egg has been laid. But at 12 hours it is only relative, and the isolation of either the anterior or the posterior fifth of the egg is still compatible with the development of a normal embryo. Interesting but intricate processes of limited regulation concerning the intermediate zone have also been described.

A typical regulation is fully realized in experiments made on a greenhouse grasshopper, *Tachycines asynamorus*.[2] The operation consists in sectioning of the embryonic anlage with a glass needle, and is performed at a relatively late stage. At the end of the period of intense mitotic divisions, about 5 days after the egg is laid, the nuclei concentrate near the micropyle in a heart-shaped area. At the same period, the embryonic anlage is formed; it sinks gradually into the vitellus and attains its full extension after 36 hours. During the first half of this period, regulation may be obtained. Longitudinal sections of the posterior region provoke a healing reaction, on both sides of which an embryo is formed. All stages of duplication, affecting whole regions, or segments, or organs, situated at various levels, may be obtained, including two embryos, complete but intimately joined (fig. 64). A gradual limitation of regulation is observed in the course of development. It first appears near the first thoracic segment, in the presumptive material of which the differentiation centre has its maximum of activity, and, from there, spreads forwards and backwards.

More recently, the egg of *Tenebrio molitor*[3] has been submitted to an analogous experimental analysis. This has resulted in the assumption of a formative centre located in the pole region, but, contrary to its situation observed in *Platycnemis*, this centre is included in the embryonic region and its influence cannot be so easily dissociated. Recent data concerning *Sialis lutaria*[4] (Neuroptera) afford an interesting transition to the so-called mosaic organization of ants, beetles and flies, where all deficiencies have definite consequences. These are species where the plas-

[1] Schnetter, 1934. [2] Krause, 1934.
[3] Ewest, 1937. [4] Du Bois, 1938.

matic layer is segregated early, and its disorganization is not
followed by restoration. In our present state of knowledge, the
differences mentioned seem mainly of a chronological order.
Among them, two common dynamic features are apparent: a

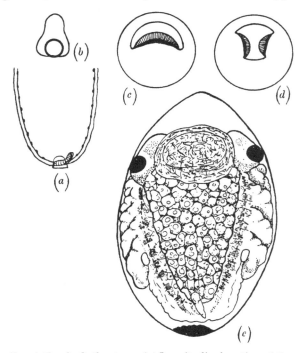

Fig. 64. Regulation in Orthoptera. (a) Longitudinal section of the posterior
region of the *Tachycines* egg, at the moment of the operation. (b) The
cordiform aspect of the egg, as seen from its posterior pole. (c) Transverse
section of an operatively divided anlage. (e) Double embryo resulting of the
operation. Redrawn from Krause, 1934.

cortical plasm polarized towards the posterior region, sometimes
with a distinct formative centre, and a considerable, largely
kinematic activity of the yolk.

CONCLUSION

The general features of the egg organization

The realization of definite forms is an almost general feature of living matter, but the early stages of egg development present it in the aspect of a specialized function, disentangled from the processes of growth and of any particular influence of the surroundings. Once the ovarial growth has been accomplished, the medium affords nothing but the mere conditions of animal, mostly aerobic, life. Moisture, temperature, the supply of O_2 and CO_2 elimination, cannot be imagined to affect the various regions of the germ in a differential, sufficiently constant manner. Gravity seems to envelop the germ with a dynamic orientated field, but, in spite of its intervention in certain experimental devices, it cannot be said to influence normal morphogenetic processes. The normal gradient results from the course of vitellogenesis, and gravity only intervenes secondarily to orientate the maturating or fertilized egg, and even then not in all species.

The main problem of Embryology is to elucidate development by the complete knowledge of the egg organization in the mature state of the germinal cell, which has in its turn to be explained by the study of oogenesis and all its antecedents. For the modern scientific mind, the causes to be considered cannot belong to any other category than the intimate physical and chemical nature of the germ materials, and their particular situation at the stage considered.

Cytological structure compels one to draw the distinction between nuclear and cytoplasmic factors. The former are at present not sufficiently analysed. Three results at least—the grafts of heterospecific cytoplasm in *Acetabularia*,[1] the injury of the cleavage nuclei in *Platycnemis*,[2] and the irradiation of the gametes in *Rana*—demonstrate the active participation of nuclear constitutents in the first morphogenetic events. Further research will have to determine whether genes, nuclear sap, or something

[1] Hämmerling, 1935.   [2] Seidel, 1932.

else are the active elements. In any case, they seem to be regularly present in all the nuclei. As genetics has induced us to imagine these as microcosms, we are inclined to believe that large possibilities of differentiation (*sensu largo*) result from reactions between the nuclear system and slight particularities of cytoplasmic areas. This plausible view still requires an actual demonstration.

Analysis of the cytoplasmic participation in morphogenesis, though really difficult, has attained substantial achievements. The data of observation are not complicated. The existence of a ground substance containing various inclusions—and especially yolk platelets accumulated opposite the nucleus—of a more condensed, viscous subcortical layer of analogous composition, and of a limiting film, may be assumed as almost general. In Coelenterates, the distinction between the superficial layers and internal material tends to be exaggerated, especially in Ctenophores. In small Chordate eggs, certain pigmentations, the yellow and grey crescents, are intimately related to morphogenesis. In *Spiralia*, pole-plasms are often cytologically visible, and, where not, seem nevertheless to be present but only demonstrable by experiment. Elsewhere, the embryologist simply observes, as in Echinoderms, a polarized big cell, with its most delicate, ultramicroscopically specialized cortical film. He has to decide whether all the facts of development may be explained on these structural bases, the interaction of nucleus and cytoplasm being admitted, or if certain supplementary properties must be attributed to the cytoplasm.

The first instance where a definite causal relation was discovered between the cytoplasm and morphogenesis, that of *Dentalium*, concerned a limited area of the germ-cell. Thence the designation of germinal localization. Such a local differentiation is also represented by the pole plasm(s) of other *Spiralia*, the posterior centre of some Insect eggs, and perhaps the genital plasm of Amphibians. But, in most of these cases, we are never sure that the described differentiation is not a mere condensation of some substance also present in the other parts of the cytoplasm. It appears thus more advisable to consider the *germinal organization in general, germinal localizations being an extreme case of condensation of some active substances.*

Thanks to a Pleiad of disinterested research workers, the

moment has now come when the material bases of germinal organization may be defined in concrete terms. In Chordates, they are the yolk gradient and the dorso-ventral field. Their quantitative interaction, coupled with their individual influence, accounts for both normal and experimental features of development. In each group, the explanation solves both the questions relative to the shape and activity of the anlagen and those concerning the experimental data of *paragenesis*. Between Orders and Classes within that large Phylum, variations only concern the quantitative relations of the two main factors. The significance of secondary, till now quite enigmatic achievements of evolution, like the foetal annexes of the Amniotes, receives a preliminary interpretation. In Echinoderms, normal and experimental data are both interpreted by the assumption of two polarized activities, the so-called animal and vegetative tendencies. A careful comparison with Chordates authorizes the deduction that the first factor is constituted by a substance mainly dispersed in the cortex of the egg, with a maximum of concentration at the animal pole. This cortical field is balanced by the vegetative gradient, the localization of which must be provisionally assumed as more general, without excluding the cortex; chemical conditions—still nearly indefinite—decrease from the vegetative pole, but at a slower rate on the presumptive ventral side of the egg. The order in which the two distinct metabolisms act by their individual properties, according to their general quantitative ratio, and by their chemical combination, according to their products, is not the same as in Chordates, but the principle is identical. In *Spiralia*, we have to do with polarized cells, possessing one or two pole-plasms. These are at first in plain continuity with the cortical film; later, they migrate inside, apparently to enter some important complex. This development suggests the combined action of one or two cortical fields with a gradient of general activity, on the same principle as in Chordates and Echinoderms. In some Insects also, two controlling centres have been recognized, one possibly cortical, the other clearly linked to internal yolk activity. A combined interaction between a "formative" field and a "differentiation" gradient may be temporarily assumed from the existing data. If we add that acoelomate eggs, and specially those of Ctenophores, also reveal the collaboration of a cortical substance and the inner cytoplasm, we may conclude that, with this conception of *a twofold chemo-differentiation distributed in a cortical field and a spatial gradient*, we have very nearly grasped the functional bases of morphogenesis, and consequently the typical features of the organization of the egg-cell as a germ.

This conception implies that the first steps of development are already a real work, in the physical meaning of the term. The energy that they utilize is used in the formation of surfaces which are endowed with very definite and important properties. *The substances produced by the reactions gradually occurring in the fertilized egg and responsible for the morphogenetic potential have a mainly cortical location.* One most interesting fact concerning the "energetics of differentiation"[1] is the slower development of half-sized eggs, observed in the sea-urchin, *Amphioxus*, the newt, *Chaetopterus*, and I may add, from personal information, in *Ascidiella*. It has also been demonstrated, in the sea-urchin, that the giant embryos obtained by fusion of fertilized eggs may develop faster than the controls.[1] The rate of oxygen consumption of two dwarf embryos from isolated blastomeres is, however, the same as that of a normal embryo—up to the gastrula stage.[2] The same probably holds true, *mutatis mutandis*, for giant germs. With an increasing volume of the system, more energy is available per unit of time and less time is required to supply the amount of energy necessary for various processes. This quite logical interpretation of Tyler[3] is capable of another, perhaps closer, expression if we consider, for the various reasons explained above, that the morphogenetic events result from a reaction between two kinds of substances, some located in the cortex, others in the whole bulk of the cell. If both concentrations are supposed to be unchanged in dwarf or in giant systems, the more rapid rate of decrease or increase of the volume relatively to the surface has evident consequences. In a larger system, for example, the increase of the inner substance available produces a proportional increase of the morphogenetic reaction products per unit of surface. The resulting acceleration should eventually be greater than is observed, were it not for the probable disadvantage

[1] Tyler, 1935.   [2] Tyler, 1933.
[3] 1935, p. 452.

resulting, for diffusion processes, from an increase of the general volume. These data concerning the energetics of development are thus quite consonant with the general interpretation arrived at in this book.[1]

When a similar analysis can be carried further and allow a closer comparison, the resemblances and differences between the germs of various Classes will appear more distinctly. Embryologists will then, in a near future, be able to define what were the long-part mutations concerning the prime chemo-differentiations and responsible for the separation of the main phyla. The ancestral relation between Worms and Molluscs was discovered by E. B. Wilson and E. G. Conklin in their pioneer study of cell lineage. The ontogenetic unity of Chordates has here been exhibited by an attentive consideration of descriptive and experimental results. An homology between the physiological cogwheels of morphogenesis has been perceived between Chordates and Echinoderms. Other achievements may reasonably be hoped for in that direction, and they will be able to give a fresh impulse to the study of Evolution.

Many details are, of course, still wanting. While the analysis must be continued and deepened on the biological level, the origin of these chemo-differentiations, and their bearing on cell physiology, has to be studied with all possible methods of physics and chemistry. Although we were here more concerned with the biological side of the problem, sufficient indications have been given of the most important results already attained by these methods.

It is important, for the successful development of this huge research programme, that old puzzling questions may now be declared to have no more than historical interest. Do form and function belong, or not, to distinct categories? Is regulation still an enigma? Is the germ initially isotropic or anisotropic? Is

[1] In a very interesting contribution to the comparative physiology of Ascidian development, Berrill (1935) emphasizes the fact that species with large eggs have a slower development than the small-egged forms. This difference, which would at first sight seem to contradict the above-quoted data, is actually of a quite other order. In Ascidians, it concerns eggs having different cytoplasmic constitutions, especially as regards yolk concentration. Is the retardation due to the more difficult CO_2 diffusion, according to the author's thesis, or to some other hindrance of the metabolism? New researches will probably solve this interesting problem.

development accomplished by preformation or by epigenesis? These much discussed problems can actually be answered in a few words. *No concrete argument can be advanced to separate form and function in their essence.* Both result, in the germ as everywhere, from the given organization. The quality, the quantity and the position of the substances mixed in the cell body—and its numerous surfaces—provide the physical conditions for chemical reactions, and form is nothing but a resultant of these inner functions. What makes its striking harmony is precisely its quantitative relation with the internal processes. These affect the shape according to their intensity, and the lines, surfaces and volumes have the intense beauty of mathematical curves.

Regulation, in the sense given to this term throughout this study, is simply a fact, which is neither more, nor less, enigmatic than any other biological occurrence. It affords indeed one of our best opportunities for becoming aware of the integral germ organization. It seems to me clear that, in all sufficiently explored cases, a satisfactory representation of this organization is available. *Regulation is thereby explained without any special hypothesis.* Distinctions such as those of presumptive potency and presumptive significance, which had their utility for a first approach of the problem, had better now be *abolished*, for they do not correspond to any intrinsic reality.[1] The development of an isolated part may be considered as the mere consequence of its initial constitution. In most cases, no extraordinary process takes place in these abnormal conditions. If isolation is able to initiate considerable changes relative to normal development— *paragenesis* contra *normogenesis*—they are only the automatic results of the previously given organization. Their physico-chemical analysis has been attempted in some circumstances, and is indeed of a fundamental interest; but there is no reason to speak of latent potencies. *Neither has an "harmonic equi-potential system" ever been actually demonstrated.* In any species, regulation is always limited in time and in space. No germ is isotropic, and its polarized organization always imposes, for quantitative reasons, a limit to regulation. We are absolutely not authorized to see in the egg the illustration of a philosophical

[1] A recommendation already expressed by Wintrebert, 1935.

abstraction, the so-called harmonic equipotential system, which perhaps has no reality whatever in the world. The term "potency" has been, of course, used in this study, but with a purely descriptive connotation, for designating a capacity or ability from the morphological point of view. The same holds true for "segregation", a most useful term for pointing to the external aspect of events that stand out as landmarks in ontogenesis. In their intimate reality, these changes concern the physical or chemical constitution of the region considered, either by formation of a new compound or by reception or discharge of a definite substance, etc.

The old conflict between preformation and epigenesis has experienced, in recent years, a vivid renewal. This arose especially from Child's interesting theory. To discuss the value of that considerable contribution to the modern biological movement falls outside the scope of this book.[1] In our limited province of early egg development, it seems to me entirely out of the question that an initially isotropic cytoplasm could acquire a polarized structure, of the kind that we know necessary for an egg, under the haphazard influence of a more active local respiration or anything similar. In opposition on this point to Huxley and de Beer,[2] I do not recognize at all the logical necessity of such an assumption. In my opinion, *the organization of the egg*—or of the sperm—*prolongs that of the germ concealed in the parent soma*, and its analysis is, for the moment, a sufficiently hard task. But I admit that this organization may include a differential distribution of respiratory activators, and that these reactions may secondarily bear on morphogenetic processes, as is likely to be the case in Amphibians and Echinoderms. In that restricted sense, Child's theory has become of concrete utility for the ontogenetic problem. But it does not mean the victory of epigenesis over preformation. For a modern mind, preformation has, of course, nothing but a figurative meaning. We have, however, no better word to assert that the *prime factors responsible for morphogenesis are contained in the germ, and not imposed on it from the outside.* On the other hand, epigenesis admirably expresses the progressive building of the organs. It will probably be ob-

---

[1] Cf. Huxley and de Beer, 1934; Huxley, 1935.
[2] 1934, p. 70.

served that my present view implies some evolution in favour of epigenesis. This is only true in the sense that the assumption of too static, restricted germinal localizations has been largely replaced by the more exact one of general fields and gradients. All recent results agree, indeed, in pointing to the appropriateness of such an interpretation, recommended by P. Weiss and so largely applicable, as J. S. Huxley[1] rightly emphasizes, to all biological processes. Concerning the egg, this modern point of view becomes particularly fertile when the fundamental duality of the morphogenetic bases and their separate location in cortex and cytoplasm has been recognized. It affords the possibility of bringing together normal and experimental data, earlier and later processes, and thus many facts which till now seemed quite disconnected. This is, I dare say, a pragmatic testimony in favour of this new orientation. But it clearly preserves the importance of the previous organization and excludes an adhesion to a purely epigenetic theory.

The way in which cytoplasmic organization is established during oogenesis, how it is affected by maturation and modified by fertilization, needs, as so many other points, further investigation. As was indicated concerning Amphibians (p. 101), Ascidians (p. 122) and Echinoderms (p 142), the existing evidence clearly suggests a bilateral organization of the unfertilized egg. Two last remarks will perhaps be of some use for a better comprehension of this initial and fascinating stage. A rôle seems to be played by the nuclear sap which, at the rupture of the germinal vesicle, partly becomes mixed with the cytoplasm, partly diffuses into the cortical layer, and often affects the permeability and the ability to be fertilized. On the other hand, attention must be paid to a probable relation between the morphogenetic activity of the superficial material, and the actual reaction of fertilization. The comparative study of this last process demonstrates, in my opinion, that the inertness of the mature but unfertilized egg is due to some inhibiting condition.[2] I doubt whether initiation of development and of morphogenesis could be brought about by different mechanisms. I also consider that, from the beginning, a definite condition is provided for the future appearance of the germ-cells and later becomes responsible for the meiotic

[1] 1935.   [2] Dalcq, Pasteels and J. Brachet, 1936.

division of these. I finally notice that yolk gradient and cortical field, or similar devices, are present but inert as long as the egg is not fertilized. Thence, the following hypothesis: fertilization or parthenogenetic treatment modifies the cortical inhibiting substance by a sort of splitting; this change corrects the deviation of the metabolism, provides the active substance of the peripheral field, and leaves a residual part, which is the genital determinant or its precursor. This triple view of the germ awakening binds together maturation, fertilization, morphogenesis, segregation of soma and germ; in a word, the whole cycle of individual life. It is inspired by the same striving for unity, which has prevailed throughout this book.

If some progress has been due to that leading idea, a large part of the recent achievements may also be attributed to the progressive introduction of quantitative concepts in causal Embryology. Hörstadius' dissociation of the sea-urchin egg into its discs; his grafts of micromeres in various numbers; Lopaschov's combination of chordo-mesoblastic pieces; Pasteels's evaluation of the interactions between cortical field and yolk gradient; my own measurements of the size attained by Ascidian organs, by the *Discoglossus* otic vesicles; my numerations of the nephrons; the interpretation of segregation by the ratio and product of the two morphogenetic elements: all these investigations have their basis in a quantitative approach. Their mathematical apparatus is still of a very elementary nature. But they attest that causal Embryology is attaining its maturity.

BIBLIOGRAPHICAL INDEX

An extensive list of publications relative to the problems discussed here will be found in the following books.

Brachet, A. 1935. *Traité d'Embryologie des Vertébrés*, 2nd edition, reviewed by A. Dalcq and P. Gérard. (Paris, Masson.)
Dalcq, A. 1935. *L'Organisation de l'œuf chez les Chordés*. (Paris, Gauthier-Villars.)
Huxley, J. S. and de Beer, G. R. 1934. *The elements of experimental Embryology*. (Cambridge University Press.)
Schleip, W. 1929. *Die Determination der Primitiventwicklung*. (Leipzig, Akademische Verlagsgesellschaft.)
Spemann, H. 1936. *Experimentelle Beiträge zu einer Theorie der Entwicklung*. (Berlin, Springer.)

For more recent contributions, the following works may be referred to. The ones indicated with an asterisk contain a considerable number of references.

*****Adelmann, A. B.** 1936. The problem of cyclopia. *Quart. Rev. Biol.* **11**, p. 161.
Balinsky, B. I. 1932. Interaction of two heteropolar equipotential systems studied by the method of complantation of Morula of the sea-urchin *Strongylocentrotus droebachiensis* (preliminary report). *Journ. Cycle Biol. Zool. Acad. Sci. Ukraine*, nos. 1–2.
Baltzer, F. and de Roche, Z. 1936. Über die Entwicklungsfähigkeit haploïder Triton alpestriskeime und über die Aufhebung der Entwicklungshemmung bei Geweben letaler bastarmerogonischer Kombination durch Transplantation in einen normalen Wirt. *Rev. Suisse Zool.* **43**, p. 495.
Bertalanffy, L. v. and Woodger, J. H. 1933. *Modern Theories of Development*. (Oxford University Press.)
*****Bounoure, L.** 1934. Recherches sur la lignée germinale chez la Grenouille rousse aux premiers stades du développement. *Ann. Sci. Nat. Zool.* 10ème série, **17**, p. 69.
—— 1937. Le sort de la lignée germinale chez la Grenouille rousse après l'action des rayons ultra-violets sur le pôle inférieur de l'œuf. *C.R. Acad. Sci.* **204**, p. 1837.
*****Brachet, J.** 1934. Étude du métabolisme de l'œuf de Grenouille (*Rana fusca*) au cours du développement. 1. La respiration et la glycolyse de la segmentation à l'éclosion. *Arch. Biol.* **45**, p. 611.
—— 1936. Le métabolisme respiratoire du centre organisateur de l'œuf de Discoglosse. *C.R. Soc. Biol.* **122**, p. 108.
*****Brachet, J. and Needham, J.** 1935. Étude du métabolisme de l'œuf de Grenouille (*Rana fusca*) au cours du développement. 4. La teneur en glycogène de l'œuf de la segmentation à l'éclosion. *Arch. Biol.* **46**, p. 821.
Brachet, J. and Shapiro, H. 1937. The relative consumption of dorsal and ventral regions of intact amphibian gastrulae, including observations on unfertilized eggs. *J. Cell. comp. Physiol.* **10**, p. 133.

186   BIBLIOGRAPHICAL INDEX

Butler, E. 1935. The developmental capacity of regions of the unincubated chick blastoderm as tested in chorio-allantoic grafts. *Journ. exp. Zool.* **70**, p. 357.

*****Child, C. M.** 1937. Differential reduction of vital dyes in the early development of Echinoderms. *Arch. Entw. Mech.* **135**, p. 426.

***** —— 1937. A contribution to the physiology of exogastrulation in Echinoderms. *Arch. Entw. Mech.* **135**, p. 457.

Cohen, A. and Berrill, N. J. 1936. The development of isolated blastomeres of the Ascidian egg. *Journ. exp. Zool.* **74**, p. 91.

—— 1936. The early development of Ascidian eggs. *Biol. Bull.* **70**, p. 78.

*****Conklin, E. G.** 1933. The development of isolated and partially separated blastomeres of *Amphioxus*. *Journ. exp. Zool.* **64**, p. 303.

*****Dalcq, A.** 1932. Étude des localisations germinales dans l'œuf vierge d'Ascidie par des expériences de mérogonie. *Arch. Anat. micr.* **28**, p. 224.

***** —— 1935. La régulation dans le germe et son interprétation. *C.R. Soc. Biol.* **119**, p. 1421.

—— 1936. L'hypothèse du "Centre initiateur somatique" (Wintrebert) et les données des translocations équatoriales chez le Discoglosse. *C.R. Soc. Biol.* **123**, p. 316.

—— 1936. L'organisation morphogénétique de l'œuf des Amphibiens aux stades blastula et gastrula. *C.R. Soc. Biol.* **123**, p. 319.

—— 1937 a. Sur l'existence de foyers pronéphritiques chez le Discoglosse. *Bull. Acad. Roy. Méd. Belg.* 6ème série, **1**, p. 495.

—— 1937 b. Sur l'existence de foyers pronéphritiques et d'autres "prédispositions" organogènes dans la blastula et la jeune gastrula du Discoglosse. *C.R. Soc. Biol.* **124**, p. 833.

—— 1937 c. Les plans d'ébauches chez les Vertébrés et la signification morphologique des annexes fœtales. *Ann. Soc. Roy. Zool. Belg.* **68**, p. 69.

*****Dalcq, A. and Pasteels, J.** 1937. Une conception nouvelle des bases physiologiques de la morphogénèse. *Arch. Biol.* **48**, p. 669.

Dalcq, A., Pasteels, J. and Brachet, J. 1936. Données nouvelles (*Asterias glacialis, Phascolion Strombi, Rana fusca*) et considérations théoriques sur l'inertie de l'œuf vierge. *Mém. Mus. Roy. Hist. nat. Belg.* 2ème série, F. 3, p. 881.

Dalcq, A. and Vandebroek, G. 1937. On the meaning of the "polar spot" in virgin and recently fertilized Ascidian eggs. *Biol. Bull.* **72**, p. 311.

Dan, K., Yanagita, T. and Sagiyama, M. 1937. Behavior of the cell surface during cleavage. *Protoplasma*, **38**, p. 66.

Dettlaff, T. 1936. Untersuchungen über das die Nervensystemanlage bei Anuren bildende Material, in Zusammenhang mit der Frage über die Wirkung des Organisators. *Zool. Jahrb. Abt. allg. Zool. Phys.* **57**, p. 203.

Ewest, A. 1937. Struktur und erste Differenzierung im Ei des Mehlkäfers *Tenebrio molitor*. *Arch. Entw. Mech.* **135**, p. 689.

Foerster, M. and Örström, A. 1933. Observations sur la prédétermination de la partie ventrale dans l'œuf d'Oursin. *Trav. Stat. Biol. Roscoff*, **11**, p. 63.

Gräper, L. 1937. Die Gastrulation nach Zeitaufnahmen linear markierter Hühnerkeime. *Anat. Anz.* (Erganzh.), **83**, p. 6.

*****Hadorn, E.** 1932. Über Organentwicklung und histologische Differenzierung in transplantierten merogonischen Bastardgeweben. *Arch. Entw. Mech.* **125**, p. 495.

—— 1937. Die entwicklungsphysiologische Auswirkung der disharmonischen Kern-Plasmakombination beim Bastardmerogen *Triton palmatus* ♀ × *Triton cristatus* ♂. *Arch. Entw. Mech.* **136**, p. 400.

Hall, E. K. 1937. Regional differences in the action of the organization centre. *Arch. Entw. Mech.* **135**, p. 671.

Halter, S. 1937. La morphogénèse de la face, étudiée comparativement chez trois Anoures. *Ann. Soc. Roy. Zool. Belg.* **67**, p. 75.

***Hämmerling, J.** 1935. Über Genomwirkungen und Formbildungsfähigkeit bei *Acetabularia. Arch. Entw. Mech.* **132**, p. 434.

***Hatt, P.** 1934. Expériences d'induction sur la gastrula de *Triton* au moyen de parties de blastoderme jeune de poulet. *Arch. Anat. micr.* **30**, p. 131.

Heatley, N. G. and Lindahl, P. E. 1937. Studies on the nature of the Amphibian organization centre. V. The nature and distribution of glycogen in the Amphibian embryo. *Proc. Roy. Soc.* B, **122**, p. 395.

Heatley, N. G., Waddington, C. H. and Needham, J. 1937. Studies on the nature of the Amphibian organization centre. VI. Inductions by the evocator-glycogen complex in intact embryos and in ectoderm removed from the individuation field. *Proc. Roy. Soc.* B, **122**, p. 403.

***Holtfreter, J.** 1935. Morphologische Beeinflussung von Urodelenektoderm bei xenoplastischer Transplantation. *Arch. Entw. Mech.* **133**, p. 367.

* —— 1935. Über das Verhalten von Anurenektoderm in Urodelen keimen. *Arch. Entw. Mech.* **133**, p. 427.

* —— 1936. Regionale Induktionen in xenoplastisch zusammen gesetzten Explantaten. *Arch. Entw. Mech.* **134**, p. 466.

***Hörstadius, Sv.** 1935. Über die Determination im Verlaufe der Eiachse bei Seeigeln. *Publ. Staz. zool. Napoli*, **14**, p. 251.

—— 1936 a. Determination in the early development of the sea-urchin. *Coll. Net.* **11**, p. 236.

* —— 1936 b. Über die zeitliche Determination im Keim von *Paracentrotus lividus* Lk. *Arch. Entw. Mech.* **135**, p. 1.

* —— 1936 c. Weitere Studien über die Determination im Verlaufe der Eiachse bei Seeigeln. *Arch. Entw. Mech.* **135**, p. 40.

* —— 1937 a. Investigations as to the localization of the micromere-, the skeleton- and the entoderm-forming material in the unfertilized egg of *Arbacia punctulata. Biol. Bull.* **73**, p. 295.

—— 1937 b. Experiments on determination in the early development of *Cerebratulus lacteus. Biol. Bull.* **73**, p. 317.

***Hörstadius, Sv. and Wolsky, A.** 1936. Studien über die Determination der Bilateralsymmetrie des jungen Seeigelkeimes. *Arch. Entw. Mech.* **135**, p. 69.

Hunt, T. E. 1937. The development of gut and its derivatives from the mesectoderm and mesentoderm of early chick blastoderms. *Anat. Rec.* **68**, p. 349.

***Huxley, J. S.** 1935. The field concept in biology. *Transactions on the Dynamics of Development*, **10**, p. 269.

***Jolly, J. and Ferester-Tadié, M.** 1936. Recherches sur l'œuf du rat et de la souris. *Arch. Anat. micr.* **32**, p. 323.

Jolly, J. and Lieure, C. 1937. Sur la culture des œufs de Mammifères. Les conditions de la culture des œufs de cobaye et de rat. *C.R. Soc. Biol.* **124**, p. 312.

***Kitchin, I. C.** 1936. Regulation und Materialverwendung bei Duplicitas cruciata. *Arch. Entw. Mech.* **134**, p. 224.

***Krause, G.** 1934. Analyse erster Differenzierungsprozesse im Keim der Gewächshausheuschrecke durch künstlich erzeugte Zwillings-Doppel- und Mehrfachbildungen. *Arch. Entw. Mech.* **132**, p. 115.

***Lehmann, F. E.** 1937. Mesodermisierung des präsumptiven Chorda-materials durch Einwirkung von Lithiumchlorid auf die Gastrula von *Triton alpestris. Arch. Entw. Mech.* **136**, p. 112.

***Lindahl, P. E.** 1936. Zur experimentellen Analyse der Determination der Dorsoventralachse beim Seeigelkeim. I. Versuche mit gestreckten Eiern. *Arch. Entw. Mech.* **127**, p. 300.

* —— 1932. Zur experimentellen Analyse der Determination der Dorso-

188   BIBLIOGRAPHICAL INDEX

ventralachse beim Seeigelkeim. II. Versuche mit zentrifugierten Eiern. *Arch. Entw. Mech.* **127**, 323.

Lindahl, P. E. 1936. Zur Kenntnis der physiologischen Grundlagen der Determination im Seeigelkeim. *Acta zool.* **17**, p. 179.

Lindahl, P. E. and Stordal, A. 1937. Zur Kenntnis des vegetativen Stoffwechsels im Seeigelei. *Arch. Entw. Mech.* **136**, p. 44.

—— —— 1937. Über die Determination der Richtung der ersten Furche im Seeigelei. *Arch. Entw. Mech.* **136**, p. 286.

Lison, L. 1936. Une méthode nouvelle de reconstruction graphique perspective. *Bull. Hist. appl.* **13**, p. 357.

Lopaschov, G. 1935. Die Umgestaltung des präsumptiven Mesoderm in Hirn teile bei Tritonkeimen. *Zool. Jahrb.* **54**, p. 299.

—— 1935. Die Entwicklungsleistungen des Gastrulamesoderm in Abhängigkeit von Veränderungen seiner Masse. *Biol. Zentr.* **55**, p. 606.

—— 1937. Über die Organbildung bei nervenlosen Organismen. II. Über die Spezifität induktiver Einflüsse. *C.R. Acad. Sci. U.R.S.S.* **15**, p. 283.

***Luther, W.** 1936. Potenzprüfungen an isolierten Teilstücken der Forellenkeimscheibe. *Arch. Entw. Mech.* **135**, p. 359.

—— 1936. Austausch von präsumptiver Epidermis u. Medullarplatte beim Forellenkeim. *Arch. Entw. Mech.* **135**, p. 484.

***Mayer, B.** 1935. Über das Regulations- und Induktionsvermögen der halbseitigen oberen Urmundlippe von Triton. *Arch. Entw. Mech.* **133**, p. 518.

***Mehrbach, H.** 1935. Beobachtungen an der Keimscheibe des Hühnchens vor dem Erscheinen des Primitivstreifens. *Zeit. Anat. Entw. ges.* **104**, p. 635.

Montalenti, G. and Maccagno, A. M. 1935. Analisi della potenza dei primi blastomeri dell' uovo de Lampreda, *Lampetra* (*Petromyzon*) *fluviatilis* L. *Arch. Ital. Anat. Embriol.* **35**, p. 69.

Morgan, T. H. 1937. *Embryologie et Génétique* (Trad. Rostand). Paris.

—— 1937. The factors locating the first cleavage plane in the egg of *Chaetopterus*. *Cytologia*, Fujii Jubilee Volume, p. 711.

Morita, S. 1936. Die künstliche Erzeugung von Einzelmissbildungen, von Zwillingen, Drillingen u. Mehrlingen im Hühnerei. *Anat. Anz.* **82**, p. 81.

—— 1937. Die künstliche Erzeugung von Mehrfachbildungen im Hühnerei. II. Vorläufige Mitteilung. *Anat. Anz.* **84**, p. 81.

Motomura, I. 1932. Über den Anlageplan und die Kinematik der Frühentwicklung bei Hynobius. *Sci. Rep. Tokohu Imp. Univ. Sendaï,* 4ème série, **6**.

—— 1935. Determination of the embryonic axis in the eggs of Amphibia and Echinoderms. *Sci. Rep. Tokohu Imp. Univ. Sendaï,* **10**, p. 211.

Oppenheimer, J. M. 1935. Transplantation and explantation on developing Teleosteans. *Anat. Rec.* **61**, p. 37.

***Pasteels, J.** 1934. Recherches sur la morphogénèse et le déterminisme des segmentations inégales chez les Spiralia. *Arch. Anat. micr.* **30**, p. 161.

* —— 1936. Centre organisateur et glycogénolyse. *Arch. Anat. micr.* **32**, p. 303.

* —— 1936. Sur l'inexistence d'un "centre initiateur somatique" (Wintrebert) chez les Amphibiens. *C.R. Soc. Biol.* **121**, p. 1394.

* —— 1936. Critique et contrôle expérimental des conceptions de P. Wintrebert sur la gastrulation du Discoglosse. *Arch. Biol.* **47**, p. 632.

* —— 1936. Étude sur la gastrulation des Vertébrés méroblastiques. I. Téléostéens. *Arch. Biol.* **47**, p. 205.

* —— 1937. II. Reptiles. *Arch. Biol.* **48**, p. 110.

* —— 1937. III. Oiseaux. IV. Conclusions générales. *Arch. Biol.* **48**, p. 381.

*Penners, A. 1936. Neue experimente zur Frage nach der Potenz der ventralen Keimhälfte von *Rana fusca*. *Zeitsch. wiss. Zool.* 148, p. 189.
* —— 1936. Regulation am Keim von *Tubifex rivulorum* Lam. nach Ausschaltung des ektodermalen Keimstreifs. *Zeitsch. f. Wiss. Zool.* (A), 149, p. 86.
Peter, K. 1931. Verwachsungsversuche mit isolierten Blastomeren von Seeigeln. *Arch. Entw. Mech.* 124, p. 17.
* —— 1934. Die erste Entwicklung des Chamäleons (*Chameleo vulgaris*), verglichen mit der Eidechse (Ei, Keimbildung, Furchung, Entodermbildung). *Zeit. Anat. Entw. Ges.* 103, p. 147.
Rashevsky, N. 1933. The theoretical physics of the cell as a basis for a general physico-chemical theory of organic form. *Protoplasma*, 20, p. 180.
Raven, C. P. 1935. Über assimilatorische Induktion in der dorsalen Urmundlippe der Amphibiengastrula. *Kon. Akad. Wetensch. Amsterdam*, 38, p. 1109.
* —— 1936. Zur Entwicklung der Ganglienleiste. V. Über die Differenzierung des Rumpfganglienleistenmaterials. *Arch. Entw. Mech.* 134, p. 122.
Reverberi, G. M. 1936. La segmentazione dei frammenti dell' uovo non fecondato di Ascidie. *Publ. Staz. Zool. Napoli*, 15, p. 198.
* —— 1937. Ricerche sperimentali sulla struttura dell' uovo fecondato delle Ascidie. *Commentationes*, vol. 1, no. 5, pp. 135–72.
*Runnström, J. 1928. Plasmastruktur und Determination des Eies von *Paracentrotus lividus* Lk. *Arch. Entw. Mech.* 113, p. 555.
* —— 1929. Über die Selbstdifferenzierung und Induktion bei dem Seeigelkeim. *Arch. Entw. Mech.* 117, p. 123.
* —— 1931. Zur Entwicklungsmechanik des Skelettmuster bei dem Seeigelkeim. *Arch. Entw. Mech.* 124, p. 273.
* —— 1933. Kurze Mitteilung zur Physiologie der Determination des Seeigelkeims. *Arch. Entw. Mech.* 129, p. 442.
* —— 1935. An analysis of the action of lithium on sea-urchin development. *Biol. Bull.* 68, p. 378.
Russel, E. S. 1930. *The interpretation of development and heredity.* (Oxford: Clarendon Press.)
Schechtman, A. M. 1937. Localized cortical growth as the immediate cause of cell division. *Science* (N.Y.), 85, p. 222.
Schmidt, G. A. 1937. Korrelationen bei der Entwicklung der Hörblase. *Zool. Anz.* 120, p. 155.
*Schnetter, M. 1934. Physiologische Untersuchungen über das Differenzierungszentrum in der Embryonalentwicklung der Honigbiene. *Arch. Entw. Mech.* 131, p. 285.
*Seidel, Fr. 1934. Das Differenzierungszentrum im Libellenkeim. I. Die dynamische Voraussetzungen der Determination und Regulation. *Arch. Entw. Mech.* 131, p. 135.
* —— 1935. Der Anlagenplan im Libellenei, zugleich eine Untersuchung über die allgemeinen Bedingungen für defekte Entwicklung u. Regulation bei dotterreichen Eiern. *Arch. Entw. Mech.* 132, p. 671.
*Spek, J. 1918. Die amoeboïden Bewegungen und Strömungen in den Eizellen einiger Nematoden während der Vereinigung der Vorkerne. *Arch. Entw. Mech.* 44, p. 217.
—— 1918. Oberflächungsdifferenzen als eine Ursache der Zellteilung. *Arch. Entw. Mech.* 44, p. 5.
* —— 1926. Über gesetzmässige Substanzverteilungen bei der Furchungen des Ctenophoreneies und ihre Beziehungen zu den Determinationsproblemen. *Arch. Entw. Mech.* 107, p. 54.
*Teissier, G. 1931. Étude experimentale du développement de quelques Hydraires. *Ann. Sci. Nat. Zool.* 10ème série, 14, p. 5.
*Ti Chow Tung. 1934. Recherches sur les potentialités des blastomesèr

chez *Ascidiella scabra*. Expériences de translocation, de combinaison et d'isolement de blastomères. *Arch. d'Anat. micr.* **30**, p. 381.

*****Töndury, G.** Beiträge zur Problem der Regulation und Induktion: Umkehrtransplantationen des mittleren Materialstreifens der beginnenden Gastrula von *Triton alpestris*. *Arch. Entw. Mech.* **134**, p. 1.

*****Twiesselmann, F.** 1938. Expériences de scission précoce de l'aire embryogène chez le poulet. *Arch. Biol.* **49** (in the press).

Tyler, A. 1933. On the energetics of differentiation. I. A comparison of the oxygen consumption of "half" and whole embryos of the sea-urchin. *Publ. Staz. Zool. Napoli,* **13**, p. 155.

—— 1935. II. A comparison of the rates of development of giant and of normal sea-urchin embryos. *Biol. Bull.* **68**, p. 451.

*****v. Ubisch, L.** 1936. Über die Organisation des Seeigelkeims. *Arch. Entw. Mech.* **134**, p. 599.

*****Vandebroek, G.** 1936. Les mouvements morphogénétiques au cours de la gastrulation chez *Scyllium canicula* Cu. *Arch. Biol.* **47**, p. 499.

—— 1936. Plasmabewegingen tydens de bevruchting in het ei van *Ascidia aspersa*. *Naturwetenschappelyk Tydschrift,* **18**, congres-nummer, p. 200.

Vintemberger, P. 1936. Sur le développement comparé des micromères de l'œuf de *Rana fusca* divisé en huit. (*a*) Après isolement. (*b*) Après transplantation sur un socle de cellules vitellines. *C.R. Soc. Biol.* **122**, p. 927. Strasbourg.

Waddington, C. H. 1935. The development of isolated parts of the chick blastoderm. *Journ. exp. Zool.* **71**, p. 273.

* —— 1937. Experiments on determination in the rabbit embryo. *Arch. Biol.* **48**, p. 273.

*****Waddington, C. H. and Cohen, A.** 1936. Experiments on the development of the head of the chick embryo. *Journ. exp. Biol.* **13**, p. 219.

Waddington, C. H. and Needham, J. 1935. Studies on the nature of the Amphibian organization centre. I. Chemical properties of the evocator (by Waddington, C. H., Needham, J., Nowinski, W. and Lemberg, R.). *Proc. Roy Soc.* B, **117**, p. 289. II. Induction by synthetic polycyclic hydrocarbons (by Waddington, C. H. and Needham, D. M.). *Proc. Roy. Soc.* B, **117**, p. 310.

—— —— 1936. Evocation, individuation and competence in Amphibian organizer action. *Proc. Roy. Acad. Amsterdam,* **39**, p. 887.

Waddington, C. H., Needham, J. and Brachet, J. 1936. Studies on the nature of the Amphibian organization centre. III. The activation of the evocator. *Proc. Roy. Soc.* B, **120**, p. 173.

*****Weiss, P.** 1935. The so-called organizer and the problem of the organization in Amphibian development. *Physiol. Rev.* **15**, p. 639.

*****Weissenberg, R.** 1936. Untersuchungen über den Anlageplan beim Neunaugenkeim. II. Das genauere Verhalten der Gliederung der zentralen Anlagezonen. *Anat. Anz.* **82**, p. 1.

—— 1936. Untersuchungen über den Anlageplan beim Neunaugenkeim. III. Analyse des dorsalen Teiles der Kopfdarmanlagezone. *Sitzungsber. Ges. Naturf. Freunde, Berlin,* p. 68.

*****Wetzel, R.** 1936. Primitivstreifen und Urkörper nach Störungsversuchen am 1–2 Tage bebrüteten Hühnchen. *Arch. Entw. Mech.* **134**, p. 357.

Wintrebert, P. 1935. Modalités et causes de la descente des blastomères de l'hémisphère animal, au cours de la blastulation des Amphibiens. *C.R. Soc. Biol.* **119**, p. 1042.

—— 1935. Le déterminisme de la gastrulation chez les Amphibiens. *C.R. Soc. Biol.* **119**, p. 1386.

—— 1935. Les rotations d'équilibre de l'œuf au cours du développement des Amphibiens. Classement et signification. *C.R. Soc. Biol.* **120**, p. 383.

—— 1935. Une théorie nouvelle du développement: l'épigénèse physiologique ou théorie des chaînes de fonctions. *C.R. Acad. Sci.* **200**, p. 1362.

Wintrebert, P. 1935. Valeur explicative de l'épigénèse physiologique. *C.R. Acad. Sci.* **201**, p. 740.

—— 1935. La régulation dans le germe et son interprétation. Remarques à propos du rapport de Mr A. Dalcq. *C.R. Soc. Biol.* réun. plén., **119**, p. 1466.

—— 1935. *Titres et Travaux scientifiques.* 2 (1923–1935). Hermann et Cie, Edit., Paris.

—— 1935. L'unité du développement et la naissance de l'individualité dans l'épigénèse physiologique des Amphibiens. *C.R. Acad. Sci.* **200**, p. 1432.

—— 1936. Les rotations d'équilibre et le plan d'ébauches de l'œuf de Discoglosse. *C.R. Soc. Biol.* **121**, p. 1584.

INDEX OF AUTHORS

INDEX OF SUBJECTS

Printed in the United States
By Bookmasters